No Headache Guide to Home Repair

# Refrigerator Repair Under $40

By
Douglas Emley

**New Century Publishing**
P.O. Box 9861
Fountain Valley, CA 92708

The Author, the publisher and all interested parties have used all possible care to assure that the information contained in this book is as complete and as accurate as possible. However, neither the publisher nor the author nor any interested party assumes any liability for omissions, errors, or defects in the materials, instructions or diagrams contained in this publication, and therefore are not liable for any damages including (but not limited to) personal injury, property damage or legal disputes that may result from the use of this book.

All major appliances are complex electro-mechanical devices. Personal injury or property damage may occur before, during, or after any attempt to repair an appliance. This publication is intended for individuals posessing an adequate background of technical experience. The above named parties are not responsible for an individual's judgement of his or her own technical abilities and experience, and therefore are not liable for any damages including (but not limited to) personal injury, property damage or legal disputes that may result from such judgements.

The advice and opinions offered by this publication are of a subjective nature ONLY and they are NOT to be construed as legal advice. The above named parties are not responsible for the interpretation or implementation of subjective advice and opinions, and therefore are not responsible for any damages including (but not limited to) personal injury, property damage, or legal disputes that may result from such interpretation or implementation.

The use of this publication acknowledges the understanding and acceptance of the above terms and conditions.

### *Acknowledgment*

*The author and the publisher wish to thank technical consultant, Dick Miller, for his expert advice and assistance in the compilation of technical information and procedure contained in this publication.*

©1994 Douglas Emley
**Published by New Century Publishing**
P.O. Box 9861
Fountain Valley, CA 92708

All rights reserved. No portion of this book may be copied or reprinted without the express prior written permission of New Century Publishing, with the exception of a review of this book whereby brief passages may be quoted by the reviewer with proper credit line. Write for permission to the above address.

International Standard Book Number   1-884348-00-9

Printed in the United States of America

# Table Of Contents

**FOREWORD**

**HOW TO USE THIS BOOK**

**CHAPTER 1: SYSTEM BASICS**

1-1 HEAT FLOW AND AIR FLOW ................................................................. 1
1-2 DEFROST SYSTEM ................................................................................ 1
1-3 TEMPERATURE CONTROL .................................................................... 2
1-4 WHERE DOES THE HEAT GO? ............................................................. 2

**CHAPTER 2: SPECIAL TYPES OF REFRIGERATORS**

2-1 HOT GAS DEFROST ............................................................................... 3
2-2 NON-SELF-DEFROSTERS ..................................................................... 3
2-3 CHILL-TYPE REFRIGERATORS ............................................................. 3
2-4 "GAS" OR AMMONIA REFRIGERATORS ............................................... 3

**CHAPTER 3: GETTING STARTED**

3-1 BEFORE YOU START ............................................................................. 5
3-2 TOOLS ..................................................................................................... 7
3-3 HOW TO USE A VOM AND AMMETER ................................................. 9
   (a) TESTING VOLTAGE ............................................................................ 9
   (b) TESTING FOR CONTINUITY ............................................................. 10
   (c) AMMETERS ........................................................................................ 11
3-4 BASIC REPAIR AND SAFETY PRECAUTIONS ................................... 12

**CHAPTER 4: COMPRESSOR RUNNING, BUT REFRIGERATOR NOT COLD**

CHAPTER 4 STEP-BY-STEP OVERVIEW ................................................... 14
4-1 CONTROLS ............................................................................................ 15
4-2 CONDENSER AND CONDENSER FAN ............................................... 17
4-3 EVAPORATOR FAN ............................................................................... 21
4-4 FROST PROBLEMS .............................................................................. 24
4-5 DEFROST SYSTEM ............................................................................... 28
   (a) DEFROST TIMER ............................................................................... 28
   (b) DEFROST HEATER ........................................................................... 30
   (c) TERMINATING THERMOSTAT .......................................................... 34
   (d) DIAGNOSIS AND REPAIR ................................................................. 35
4-6 WHIRLPOOL FLEX-TRAY DEFROSTING ............................................ 39

4-7 HOT GAS DEFROST PROBLEMS ..................................................... 44
4-8 UNEVEN FROST PATTERNS OR NO FROST AT ALL ................................. 45
4-9 COLD CONTROL ................................................................ 45

## CHAPTER 5: COMPRESSOR NOT RUNNING, REFRIGERATOR NOT COLD

CHAPTER 5 STEP-BY-STEP OVERVIEW ................................................. 48
5-1 POWER ....................................................................... 49
5-2 CONTROLS .................................................................... 49
5-3 DIAGNOSIS AND REPAIR ........................................................ 49
   (a) DEFROST TIMER ........................................................... 49
   (b) POWER TO COMPRESSOR ..................................................... 51
   (c) COLD CONTROL ............................................................ 51
   (d) WIRES AND ELECTRICAL .................................................... 52
   (e) ELECTRICAL / MOTOR STARTING ............................................. 52

## CHAPTER 6: ICE OR WATER BUILDUP

6-1 DEFROST DRAIN SYSTEM ....................................................... 55
6-2 DIAGNOSIS AND REPAIR ....................................................... 57
6-3 DRAIN PAN MULLION HEATERS .................................................. 59

## CHAPTER 7: FLUKES AND UNUSUAL COMPLAINTS

7-1 KID CAPERS .................................................................. 61
7-2 THE HOLE-IN-THE-WALL GANG ................................................... 61
7-3 DOOR SEALS AND ALIGNMENT ................................................... 62
7-4 MOVING DAY .................................................................. 63
7-5 A SHOCKING EXPERIENCE (MULLION HEATERS) ................................... 63
7-6 MICE CAPADES ................................................................ 64
7-7 ICEMAKERS AND IN-DOOR WATER DISPENSERS .................................... 64
7-8 SEE THE LIGHT ............................................................... 64
7-9 BAD ODORS ................................................................... 65
7-10 FIX THE LITTLE STUFF ....................................................... 65
7-11 STRANGE NOISES ............................................................. 65
7-12 FIRE IN THE FRIDGE! ........................................................ 65

## INDEX

# FOREWORD

## *WHAT THIS BOOK WILL DO FOR YOU*
(and what it won't!)

This book **will** tell you how to fix your vapor-compression domestic (home) refrigerator and freezer units. (This represents 99+ percent of the home refrigerators in service today.)

This book **will not** tell you how to fix your gas refrigerator, home air conditioner, industrial or commercial or any very large air conditioning or refrigeration unit. The support and control systems for such units are usually very similar in function to those of smaller units, but vastly different in design, service and repair.

We **will** show you the easiest and/or fastest method of diagnosing and repairing your refrigerator.

We **will not necessarily** show you the cheapest way of doing something. Sometimes, when the cost of a part is just a few dollars, we advocate replacing the part rather than rebuilding it. We also sometimes advocate replacement of an inexpensive part, whether it's good or bad, as a simplified method of diagnosis or as a preventive measure.

We **will** use only the simplest of tools; tools that a well-equipped home mechanic is likely to have and to know how to use, including a VOM.

We **will not** advocate your buying several hundred dollars' worth of exotic equipment or getting advanced technical training to make a one-time repair. It will usually cost you less to have a professional perform this type of repair. Such repairs represent only a very small percentage of all needed repairs.

We **do not** discuss the vapor compression cycle, thermodynamic laws, specific heat, latent heat, sensible heat, measurement of heat, molecular theory, cutting, brazing, capillaries and expansion valves, compressor operation, condenser operation, evaporator operation, driers or Freons 11, 12, 13, 22, 113, 114, 500, or 502. There are already many very well-written textbooks on these subjects and most of them are not likely to be pertinent to the job at hand; fixing your fridge!

We **do** discuss rudimentary heat flow and airflow, frost, drainage, and simple electrical circuits.

You are expected to know some simple natural laws. For example; when you compress a gas, it warms up; warm air rises by convection; air usually has humidity; water can be found in a solid state (ice), a liquid state, or a gaseous state (steam or humidity).

You should know how to cut, strip, and splice wire with crimp-on connectors, wire nuts and electrical tape. You should know how to measure voltage and how to test for continuity with a VOM (Volt-Ohm Meter). If you have an ammeter, you should know how and where to measure the current in amps. If you don't know how to use these meters, there's a brief course on how to use them (for *our* purposes *only*) in section 3-3. See section 3-2 before you buy either or both of these meters.

A given procedure was only included in this book if it passed the following criteria:
1) The job is something that the average couch potato can complete in one afternoon, with no prior knowledge of the machine, with tools a normal home handyman is likely to have.
2) The parts and/or special tools required to complete the job are easily found and not too expensive.
3) The problem is a common one; occuring more frequently than just one out of a hundred machines.

Certain repairs which may cost more than $40 may be included in this book, if they pass the following criteria:
1) The cost of the repair is still far less than replacing the machine or calling a professional service technician, and
2) The repair is likely to yield a machine that will operate satisfactorily for several more years, or at least long enough to justify the cost.

In certain parts of the book, the author expresses an opinion as to whether the current value of a particular machine warrants making the repair or ''scrapping'' the machine. Such opinions are to be construed as opinions ONLY and they are NOT to be construed as legal advice. The decision as to whether to take a particular machine out of service depends on a number of factors that the author cannot possibly know and has no control over; therefore, the responsibility for such a decision rests solely with the person making the decision.

I'm sure that a physicist or a thermodynamicist reading this book could have a lot of fun tearing it apart because of my deliberate avoidance and misuse of technical terms. However, this manual is written to simplify the material and inform the novice, not to appease the scientist.

*NOTE: The diagnosis and repair procedures in this manual do not necessarily apply to brand-new units, newly-installed units or recently relocated units. Although they **may** posess the problems described in this manual, refrigerators that have recently been installed or moved are subject to special considerations not taken into account in this manual for the sake of simplicity. Such special considerations include installation parameters, installation location, the possibility of manufacturing or construction defects, damage in transit, and others.*

*This manual was designed to assist the novice technician in the repair of home (domestic) refrigerators that have been operating successfully for an extended period of months or years and have only recently stopped operating properly, with no major change in installation parameters or location.*

*There is only one exception to this rule; see section 7-4.*

# HOW TO USE THIS BOOK

**STEP 1: PLEASE READ THE DISCLAIMER LOCATED ON THE COPYRIGHT PAGE.** This book is intended for use by people who have a little bit of mechanical experience or aptitude, and just need a little coaching when it comes to appliances. If you don't fit that category, you may want to rethink trying to fix it yourself! We're all bloomin' lawyers these days, y'know? If you break something or hurt yourself, no one is responsible but **you**; not the author, the publisher, the guy or the store who sold you this book, or anyone else. If you don't understand the disclaimer, get a lawyer to translate it **before** you start working.

Read the safety and repair precautions in section 3-4. These should help you avoid making any *really* bad mistakes.

**STEP 2: READ CHAPTERS 1, 2 & 3:** Know what kind of refrigerator you have and basically how it works. When you go to the appliance parts dealer, have the nameplate information at hand. Have the proper tools at hand, and know how to use them.

**STEP 3: SCAN THE BLOCKS BELOW FOR YOUR SYMPTOMS.** When you find them, go to the appropriate chapter or section and follow the instructions. If you don't see your symptoms in the chart below, read through Chapter 7. It might help you figure out for yourself what's wrong.

**STEP 4: FIX THE BLOOMIN' THING!** If you can, of course. If you're just too confused, or if the book recommends calling a technician for a complex operation, call one.

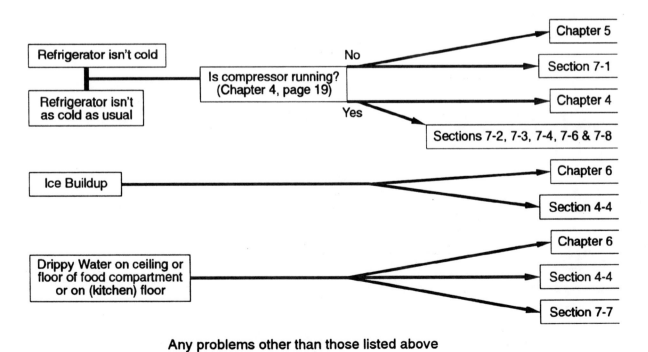

**Any problems other than those listed above
read through Chapter 7**

# Chapter 1

# SYSTEM BASICS

## 1-1. HEAT FLOW AND AIR FLOW

In thinking about the way your refrigerator keeps things cool, there are a few seat-of-the-pants thermodynamic laws that you need to consider:

1) *Don't think of your refrigerator as MAKING THINGS COLD. Think of it as REMOVING HEAT from whatever you put in it.*

2) *Remember that heat always flows* **from** *something of a* **higher** *temperature to something of a* **lower** *temperature. The greater the temperature difference, the faster the heat flow.*

3) *Heat will continue to flow* **from** *the* **warm***er object* **to** *the* **cold***er object until their temperatures are equal.*

Thus, when you put warm food into a cold refrigerator, heat starts flowing from the warm food into the cold air around it. The food starts to get cooler. (The temperature goes down.)

Since the air inside the fridge has now absorbed some heat and is warmer than it should be, we need to carry it away from the food, replace it with new cold air and remove the excess heat from it. This is why airflow is so important inside the fridge.

A fan is provided inside the fridge to continually circulate internal air. It draws in warm air from all sections of your refrigerator and blows it across the **EVAPORATOR**, or **"COOLING COILS."** This fan is called the **EVAPORATOR FAN.**

As the air flows across the *out*side of the evaporator, it gives up its excess heat to Freon circulating *in*side the evaporator. The Freon in the evaporator runs at a very cold temperature, somewhere between about 9 and 40 degrees below zero, so that the heat will flow out of the air and into the Freon very quickly.

## 1-2. DEFROST SYSTEM

However, when you opened that door to put the food into the fridge, you also let in a big charge of warm, relatively moist (humid) air. The evaporator is SO cold that the humidity from the air will freeze directly onto it, creating **FROST.** If enough frost builds up on the evaporator, air will not be able to flow across it.

To prevent too much frost from collecting on the evaporator coils, a self-defrosting refrigerator will actually stop itself for a few minutes every six to twelve hours and melt its own frost. This defrost cycle is controlled by an electric **DEFROST TIMER.** The timer stops the compressor (cooling system) and starts an electric **DEFROST HEATER** located directly beneath the evaporator coils. The heat rises and melts the frost.

The frost water is drained away, usually to the little pan that you see beneath your refrigerator in most models. Eventually it will evaporate away. If all the frost melts before the defrost timer finishes the defrost cycle, a **TERMINATING THERMOSTAT** will keep the defrost heater from overheating the evaporator compartment. It is wired in series with the heater. When the compartment reaches a certain temperature, the terminating thermostat will open and shut off the defrost heater.

## 1-3. TEMPERATURE CONTROL

As the food in the fridge gets colder, it gives off less heat, and the air inside the fridge will remain colder. A thermostat called a **COLD CONTROL** will cycle the cooling system on and off to keep the temperature inside your fridge within a certain range. You can adjust that range using one of the dials within your fridge.

On most fridges, all the cold air for both the food compartment and the freezer compartment is produced in one evaporator. Since the freezer is so much colder than the food compartment, most of the cold air that is produced circulates to the freezer compartment. Only a small amount is needed in the food compartment to keep it down to the proper temperature. This amount is adjusted by a small **AIR DOOR** in the duct between the evaporator and the food compartment. The control for this air door is the other of the two dials within your fridge.

## 1-4. WHERE DOES THE HEAT GO?

In order to re-use the Freon for cooling more air, the heat that WAS is your food and is NOW in your Freon must somehow be gotten rid of. The Freon gas goes to the **COMPRESSOR** where it is compressed into a hot gas (remember, when you compress a gas, it gets warmer.) The Freon then flows through the **CONDENSER**, which is the warm grille you'll find either behind or underneath your refrigerator. It resembles the radiator in your car and acts in much the same manner. The heat from your food, plus whatever heat was added during compression, will flow *from* the hot Freon in your condenser *to* the cooler air in your kitchen.

In some models a **CONDENSER FAN** is fitted to circulate air across the condenser. Air circulation is very important to heat flow. On some models, the warmth from the condenser is used to evaporate the defrost water. That's also the source of the warm air you might feel blowing out from beneath your fridge while it is running.

After the Freon loses the excess heat in the condenser, it is de-compressed, or expanded, by simply putting it through a constriction that lets the pressure drop. This readies the Freon to go through the evaporator again, and the cycle begins all over.

# Chapter 2

# SPECIAL TYPES OF REFRIGERATORS

A vast majority of the refrigerators built within the last 30-40 years (and thus most likely to still be in service) were built using the systems described in Chapter 1; however, there are a few exceptions:

## 2-1. HOT GAS DEFROST

Certain refrigerators use hot Freon circulated backwards through the evaporator to melt frost, rather than an electric heater. Though there are more recent models, these are generally built in the 1960's and before and represent a very small minority of the domestic refrigerators still in service.

## 2-2. NON SELF-DEFROSTERS

These units do not have a timer that initiates the defrost mode. Some must be initiated manually (usually there is an obvious defrost button) and some have no defrost at all--they must be unloaded, unplugged, and let defrost by themselves.

## 2-3. CHILL-TYPE REFRIGERATORS

These refrigerators have a very cold plate or coils (tubes) located near the top of the cold compartment. There is no evaporator fan and air circulates naturally. By convection, warm air rises to the top of the compartment where the coils are, and the cold air sinks towards the bottom.

There may be separate plates in the freezer and refrigerator sections. Often these refrigerators have no defrost systems; if they do, it's usually a hot-gas defrost system. Most "micro-" or "mini-" fridges are chill-type. Most freezer (only) units have no evaporator fans. Generally these units have no condenser fan either; condenser air circulation is also by convection.

## 2-4. GAS REFRIGERATORS (AMMONIA SYSTEMS)

IF YOU HAVE A "GAS" REFRIGERATOR, NOTHING IN THIS BOOK APPLIES TO YOUR UNIT. It uses an entirely different operating system; it is NOT a vapor-compression-cycle unit. These are generally found in RV and yacht installations, where electricity is frequently unavailable for extended periods of time.

If you suspect that you may have any of these units, ask your appliance parts retailer or dealer to confirm your suspicions. (See Section 3-1; "BEFORE YOU START.") With a make, a model number and perhaps a serial number, he should be able to tell you what you have.

# Chapter 3

# DIAGNOSIS AND REPAIR BASICS

## 3-1. BEFORE YOU START

Find yourself a good appliance parts dealer. You can find them in the Yellow Pages under the following headings:

- APPLIANCES, HOUSEHOLD, MAJOR
- APPLIANCES, PARTS AND SUPPLIES
- REFRIGERATORS, DOMESTIC
- APPLIANCES, HOUSEHOLD, REPAIR AND SERVICE

Call a few of them and ask if they are a repair service, or if they sell parts, or both. Ask them if they offer free advice with the parts they sell. (Occasionally, stores that offer both parts and service will not want to give you advice.) Often, the parts counter men are ex-technicians who got tired of the pressures of going into stressed-out peoples' houses and fixing appliances. They can be your best friends; however, you don't want to badger them with TOO many questions, so know your basics before you start asking questions.

*Some* parts houses may offer service too. Be careful! They may try to talk you out of even *trying* to fix your own refrigerator. They'll tell you it's too complicated, then in the same breath, "guide" you to their service department. Who are you gonna believe, me or them? Not all service/parts places are this way, however. If they genuinely *try* to help you fix it yourself and you find that you can't fix the problem, they may be a really good place to look for service.

When you go into the store, have ready your make, model and serial number from the *nameplate* of the fridge (not from some sticker inside the fridge). If there is a B/M number on the nameplate, have that with you, too.

## NAMEPLATE INFORMATION

The metal nameplate information is usually found in one of the places shown in Figure 1:

A) Along the bottom panel; left, right or anywhere in-between.

B) Inside the fridge or freezer section, near the bottom. You may have to remove a crisper drawer to see it.

C) Remove the kickplate and look along the condenser air openings.

D) Somewhere on the back of the refrigerator, usually very high or very low, or possibly on any wiring diagram that may be pasted to the back of the refrigerator.

E) If you absolutely cannot find a metal nameplate, some refrigerators have a paper sales sticker left on, just inside the door. This will be an incomplete model number, but it is better than nothing and it should be good enough to get most parts with.

If all else fails, check the original papers that came with your fridge when it was new. They should contain the model number SOMEWHERE.

If you have absolutely NO information about the fridge anywhere, make sure you bring your old part to the parts store with you. Sometimes they can match an old part by looks or by part number.

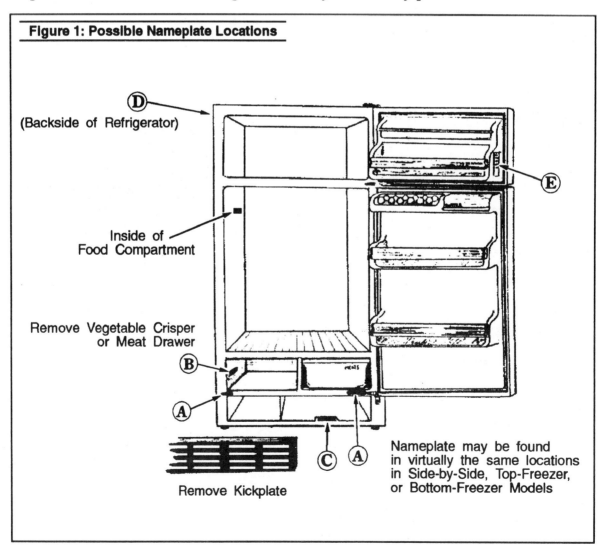

Figure 1: Possible Nameplate Locations

## 3-2. TOOLS (Figure 2)

The tools that you will probably need are listed below. Some are optional. The reason for the option is explained.

SCREWDRIVERS: Both flat and phillips head; two or three different sizes of each. It is best to have at least a stubby, 4" and 6" sizes.

NUTDRIVERS: You will need at least a 1/4" and a 5/16" nut driver. 4 or 6" ones should suffice, but it's better to have stubbies, too.

ELECTRICAL PLIERS or STRIPPERS and DIAGONAL CUTTING PLIERS: For cutting and stripping small electrical wire.

ALLIGATOR JUMPERS (sometimes called a "CHEATER" or "CHEATER WIRE"): small gauge (14-16 gauge or so) and about 12 to 18 inches long; for testing electrical circuits. Available at your local electronics store. Cost: a few bucks for 4 or 5 of them.

VOM (VOLT-OHM METER): For testing circuits. If you do not have one, get one. An inexpensive one will suffice, as long as it has both "A.C. Voltage" and "Resistance" (i.e. R x 1, R x 10, etc.) settings on the dial. It will do for our purposes. If you are inexperienced in using one, get an analog (pointer) type (as opposed to a digital.)

FLASHLIGHT: For obvious reasons.

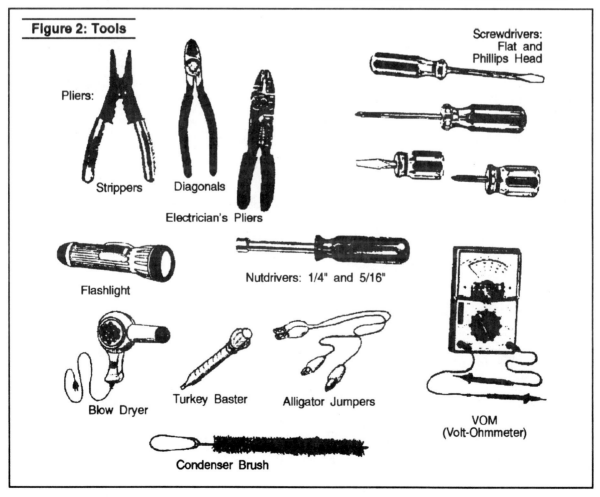

Figure 2: Tools

CONDENSER BRUSH: For cleaning out that dusty "black hole" beneath your fridge, otherwise known as your condenser. It is a long, stiff-bristled brush especially made for knocking out massive wads of dust from your condenser grille. I have seen jury-rigged bottle brushes and vacuums used, neither of which clean sufficiently. C'mon--buy the right tool for the job. They're cheap enough.

BLOW DRYER and SYRINGE TYPE TURKEY BASTER: For manually melting frost and ice.

BUTT CONNECTORS, CRIMPERS, WIRE NUTS AND ELECTRICAL TAPE: For splicing small wire.

## OPTIONAL TOOLS (Figure 3)

SNAP-AROUND AMMETER: For determining if electrical components are energized. Quite useful; but a bit expensive, and there are alternative methods. If you have one, use it; otherwise, don't bother getting one.

EXTENDABLE INSPECTION MIRROR: For seeing difficult places beneath the refrigerator and behind panels.

CORDLESS POWER SCREWDRIVER OR DRILL/DRIVER WITH MAGNETIC SCREWDRIVER AND NUTDRIVER TIPS: For pulling off panels held in place by many screws. It can save you lots of time and hassle.

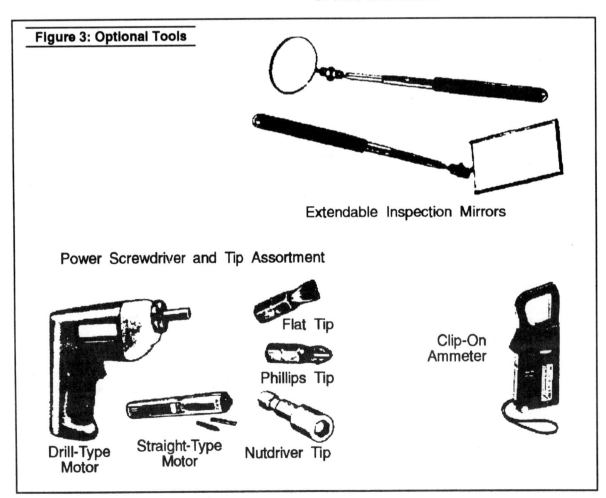

Figure 3: Optional Tools

# DIAGNOSIS AND REPAIR BASICS

## 3-3. HOW TO USE A VOM AND AMMETER

Many home handymen are very intimidated by electricity. It's true that diagnosing and repairing electrical circuits requires a *bit* more care than most operations, due to the danger of getting shocked. But there is no mystery or voodoo about the things we'll be doing. Remember the rule in section 3-4 (1); while you are working on a circuit, energize the circuit only long enough to perform whatever test you're performing, then take the power back off it to perform the repair. You need not be concerned with any theory, like what an ohm is, or what a volt is. You will only need to be able to set the VOM onto the right scale, touch the test leads to the right place and read the meter.

In using the VOM (Volt-Ohm Meter) for our purposes, the two test leads are always plugged into the "+" and "-" holes on the VOM. (Some VOMs have more than two holes.)

### 3-3(a). TESTING VOLTAGE (Figure 4)

Set the VOM's dial on the lowest VAC scale (A.C. Voltage) that's over 120 volts. For example, if there's a 50 setting and a 250 setting on the VAC dial, use the 250 scale, because 250 is the lowest setting over 120 volts.

Touch the two test leads to the two metal contacts of a live power source, like a wall outlet or the terminals of the motor that you're testing for voltage. (*Do not* **jam** *the test leads into a wall outlet!*) If you are getting power through the VOM, the meter will jump up and steady on a reading. You *may* have to convert the scale in your head. For example, if you're using the 250 volt dial setting and the meter has a "25" scale, simply divide by 10; 120 volts would be "12" on the meter.

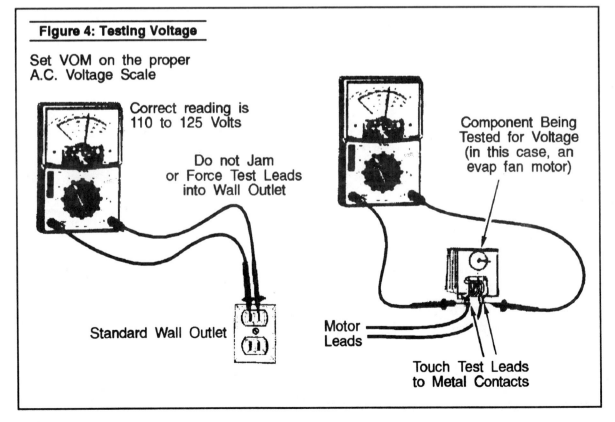

**Figure 4: Testing Voltage**

Set VOM on the proper A.C. Voltage Scale

Correct reading is 110 to 125 Volts

Do not Jam or Force Test Leads into Wall Outlet

Standard Wall Outlet

Component Being Tested for Voltage (in this case, an evap fan motor)

Motor Leads

Touch Test Leads to Metal Contacts

## 3-3(b). TESTING FOR CONTINUITY (Figure 5)

Don't let the word "continuity" scare you. It's derived from the word "continuous." In an electrical circuit, electricity has to flow *from* a power source back *to* that power source. If there is any break in the circuit, it is not continuous, and it has no continuity. "Good" continuity means that there is no break in the circuit.

For example, if you were testing a heater element to see if it was burned out, you would try putting a small amount of power through the heater. If the heater element was burned out, there would be a break in the circuit, the electricity wouldn't flow, and your meter would show no continuity.

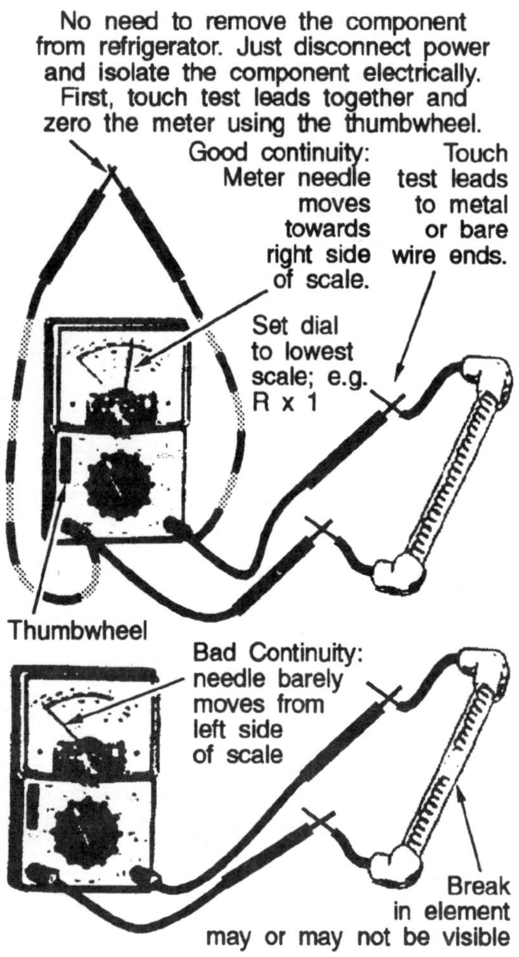

That is what the *resistance* part of your VOM does; it provides a small electrical current (using batteries within the VOM) and measures how fast the current is flowing. For our purposes, it doesn't matter how *fast* the current is flowing; only that there *is* current flow.

To use your VOM to test continuity, set the dial on (resistance) R x 1, R x 10, or whatever the *lowest* setting is. Touch the metal parts of the test leads together and read the meter. It should peg the meter all the way on the right side of the scale, towards "0" on the meter's "resistance" scale. If the meter does not read zero resistance, adjust the thumbwheel on the front of the VOM until it *does* read zero. If you cannot get the meter to read zero, the battery in the VOM is low; replace it.

If you are testing, say, a heater, first make sure that the heater leads are not connected to *anything, especially* a power source. If the heater's leads are still connected to something, you *may* get a reading through that something. If there is still live power on the item you're testing for continuity, you will burn out your VOM in microseconds and possibly shock yourself.

Touch the two test leads to the two bare wire ends or terminals of the heater. You can touch the ends of the wires and test leads with your hands if necessary to get better contact. The voltage that the VOM batteries put out is very low, and you will not be shocked. If there is NO continuity, the meter won't move. If there is GOOD continuity, the meter will move toward the right side of the scale and steady on a

**Figure 5: Testing for Continuity**

No need to remove the component from refrigerator. Just disconnect power and isolate the component electrically. First, touch test leads together and zero the meter using the thumbwheel.

Good continuity: Meter needle moves towards right side of scale.

Touch test leads to metal or bare wire ends.

Set dial to lowest scale; e.g. R x 1

Thumbwheel

Bad Continuity: needle barely moves from left side of scale

Break in element may or may not be visible

# DIAGNOSIS AND REPAIR BASICS

reading. This is the resistance reading and it doesn't concern us; we only care that we show good continuity. If the meter moves only very little and stays towards the left side of the scale, that's BAD continuity; the heater is no good. In a glass-tube or bare-element heater, you *may* be able to see the physical break in the heater element, just like you can in some light bulbs.

If you are testing a switch or a thermostat, you will show little or no resistance (good continuity) when the switch or thermostat is closed, and NO continuity when the switch is open. If you do not, the switch is bad.

## 3-3(c). AMMETERS

Ammeters are a little bit more complex to explain without going into a lot of electrical theory. If you own an ammeter, you probably already know how to use it.

If you don't, don't get one. Ammeters are expensive. And for *our* purposes, there are other ways to determine what an ammeter tests for. If you don't own one, skip this section.

For our purposes, ammeters are simply a way of testing for continuity without having to cut into the system or to disconnect power from whatever it is we're testing.

Ammeters measure the current in amps flowing through a wire. The greater the current that's flowing *through* a wire, the greater the density of the magnetic field, or *flux*, it produces *around* the wire. The ammeter simply measures the density of this flux, and thus the amount of current, flowing through the wire. To determine continuity, for our purposes, we can simply isolate the component that we're testing (so we do not accidentally measure the current going through any other components) and see if there's *any* current flow.

To use your ammeter, first make sure that it's on an appropriate scale (0 to 10 or 20 amps will do). Isolate a wire leading directly to the component you're testing. Put the ammeter loop around that wire and read the meter. (Figure 6)

What if that you have trouble finding the wire lead to, say, a defrost heater that you're testing, because the wires are frozen into a block of ice?

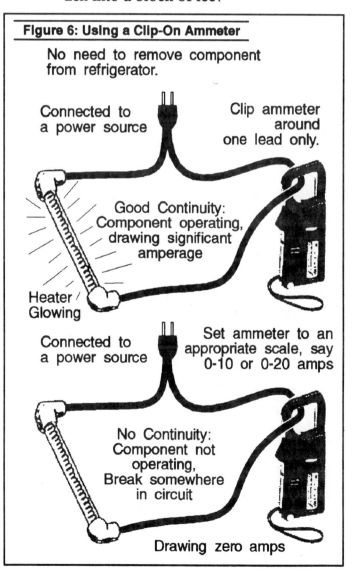

**Figure 6: Using a Clip-On Ammeter**

No need to remove component from refrigerator.

Connected to a power source

Clip ammeter around one lead only.

Good Continuity: Component operating, drawing significant amperage

Heater Glowing

Connected to a power source

Set ammeter to an appropriate scale, say 0-10 or 0-20 amps

No Continuity: Component not operating, Break somewhere in circuit

Drawing zero amps

If you isolate all systems so the heater is the only thing energized, then you can test the easily accessible main power cord for current, and it must be the current being used by the heater.

This is relatively easy to do. Turn the "energy saver" switch to the "economy" position to shut off the anti-sweat mullion heaters (See section 4-1.) Close the refrigerator door to make sure the lights are off. Set the defrost timer to the defrost mode. This will stop the compressor, and energize the defrost heater (*if* it is working; *that*, of course, is what we're testing.)

Remove the lower back panel of the fridge and find where the main power cord branches off in two directions. (See Figure 7) Test one lead (*one* lead *only*) for current flow. There may still be a tiny mullion heater energized in the butter conditioner or on the defrost drain pan, but the current that these heaters draw is negligible for our purposes (less than an amp). If you have *any* substantial current flow, it's *got* to be through the defrost heater. If you don't, the defrost heater or terminating thermostat is probably defective.

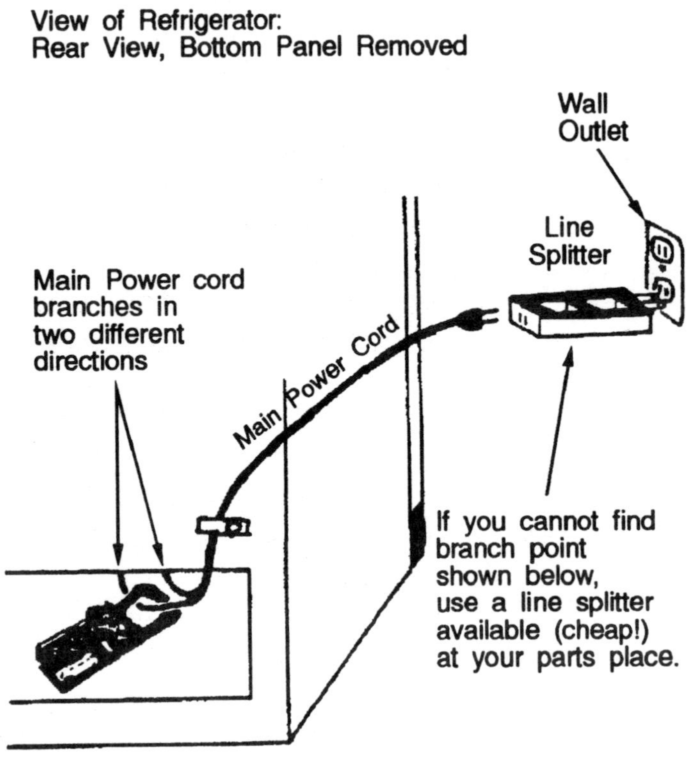

### 3-4. BASIC REPAIR AND SAFETY PRECAUTIONS

1) Always de-energize (pull the plug or trip the breaker on) any refrigerator that you're disassembling. If you need to re-energize the refrigerator to perform a test, make sure any bare wires or terminals are taped or insulated. Energize the unit only long enough to perform whatever test you're performing, then disconnect the power again.

2) **_NEVER EVER_** chip or dig out ice from around the evaporator with a sharp instrument or knife. You WILL PROBABLY puncture the evaporator and you WILL PROBABLY end up buying a new refrigerator. Use hot water and/or a blow dryer to melt ice. If you use a blow dryer, take care not to get water in it and shock yourself. Better yet, if you

**Figure 7: Testing Amps in the Main Power Line**

View of Refrigerator:
Rear View, Bottom Panel Removed

Main Power cord branches in two different directions

Main Power Cord

Wall Outlet

Line Splitter

If you cannot find branch point shown below, use a line splitter available (cheap!) at your parts place.

have the time and patience, leave the fridge open for a few hours and let the ice melt naturally. You can remove large, loose chunks of ice in the evaporator compartment by hand, but make sure there aren't any electrical wires frozen into the chunks of ice before you start pulling them.

3) Always re-install any removed duck seal, heat shields, styrofoam insulation, or panels that you remove to access *anything*. They're there for a reason.

4) You may need to empty your fridge or freezer for an operation. If you do not have another fridge (or a friend with one) to keep your food in, you can generally get by with an ice chest or a cardboard box insulated with towels for a short time. Never re-freeze meats; if they've already thawed, cook them and use them later.

5) If the manual advocates replacing the part, REPLACE IT!! You might find, say, a fan motor that has stopped for no apparent reason. Sometimes you can flip it with your finger and get it going again. The key words here are *apparent reason*. There is a reason that it stopped--you can bet on it-- and if you get it going and re-install it, you are running a very high risk that it will stop again. If *that* happens, you will have to start repairing your refrigerator *all* over again. There may be a hard spot in the bearings; or it may only act up when it is hot, or cold...there are a hundred different "what if's." Very few, if any, of the parts mentioned in this book will cost you over ten or twenty dollars. Don't be penny-wise and dollar-dumb. Replace the part.

6) Refrigerator defrost problems may take a week or more to reappear if you don't fix the problem the first time. That's how long it will take the evaporator to build up enough frost to block the airflow again. After fixing a defrost problem, keep an eye out for signs of a recurrence for at least a week. The sooner you catch it, the less ice you'll have to melt.

7) You may stop the compressor from running using the defrost timer or cold control, by cutting off the power to the fridge, or simply by pulling the plug out of the wall. However, if you try to restart it within a few minutes, it *may* not start; you will hear it buzzing and clicking. (See section 5-3(e)). If the system has not had enough time for the pressure within to equalize, there will be too much back pressure in the system for the compressor motor to overcome when trying to start. This is nothing to be alarmed about. Simply remove the power from the compressor for a few more minutes until the compressor *will* restart.

8) Do not lubricate any of the timers or motors mentioned in this manual. They are permanently self-lubricated. In a cold environment, oil will become more viscous and *increase* friction, rather than decrease it. If you have a sticky fan motor or timer, replace it.

# Chapter 4 Overview
## Step-by-Step

Complaint: Warm refrigerator, or not as cold as usual
Chapter Qualifier: Compressor is running

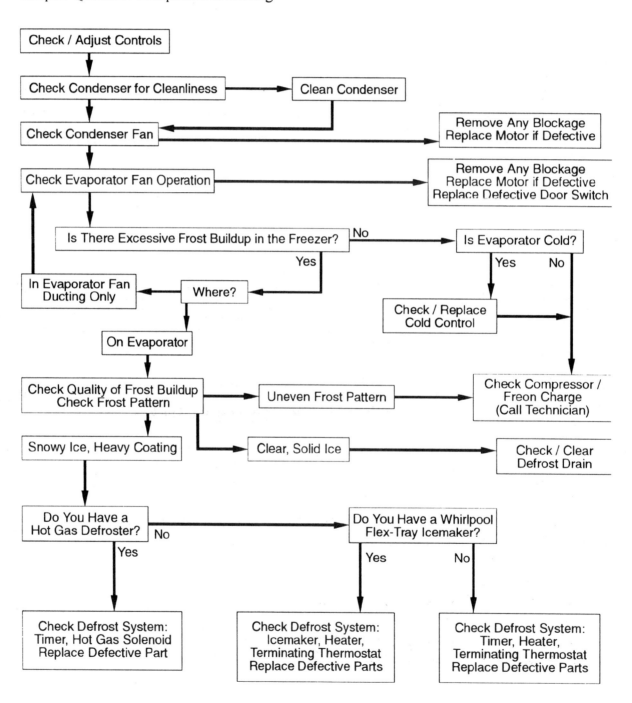

# Chapter 4

# COMPRESSOR IS RUNNING BUT REFRIGERATOR IS NOT COLD

Before you perform any of the other tests in this chapter, make sure that the compressor is running. If it is *not* running, see Chapter 5.

Some refrigerators are very quiet and smooth when they operate. If you cannot hear your refrigerator running or feel the compressor vibrating, you must investigate further.

First, try turning the cold control to the "off" position or unplugging the fridge; this will stop the compressor. Do you hear or feel a difference? If so, the compressor *was* running. WAIT SEVERAL MINUTES before turning the compressor back on for your diagnostic checks. The reason for waiting before you restart it is explained in section 3-4 (7).

If you perform the above test and do not feel a difference, try "listening with a screwdriver." Access the compressor by opening the back panel and place the metal end of a long screwdriver against the compressor and your ear against the plastic end of the screwdriver. You should hear the compressor running. If you are still unsure and you own an ammeter, test the current draw of the compressor at the compressor leads. If the compressor is running, it should draw about 6 amps.

## 4-1. CONTROLS

If your compressor is running and your refrigerator is warm in both compartments (or not as cold as usual, i.e. chilly but meats are thawing,) first check your CONTROLS. You never know if your kids got in there and messed around with them. Set them on mid-range settings. See section 7-1 on KID CAPERS for some interesting stories about this subject.

Inside either the freezer or refrigerator compartment you will generally find at least two dial type controls.

One of them, called the cold control, is an electric switch that starts and stops the compressor based on the temperature that it senses inside the compartment.

The other dial is an air door that controls the small amount of air that passes to the food compartment while the evaporator fan is running. (See Section 1-3)

Either dial may be marked with any one of a dozen different labels: "refrigerator control," "freezer control," "food compartment control," etc. Determining which is which can get a bit confusing. If the knob has an "off" setting which stops the compressor from running, it is the *cold control*.

In the absence of an "off" setting, the easiest way to tell them apart is to pull the plastic knob off the control. The *cold control* will have a wide tang and a narrow tang. (See Figure 8) The *air door* will usually have a plastic or metal "D"-shaped shaft (a round shaft with a flat) to which it attaches, although this is not always the case.

If one dial is in the freezer section and one is in the food section, the one in the *freezer* section is the *cold control*, and the one in the *food* section is the *air door*.

If the knobs will not come off with a firm pull, or you are still unsure of which control is which, try putting your hand in front of the air vents in the food compartment and manipulating the controls. Make sure the evap fan is running''you *may* have to *tape* the door switch so it stays on. If you are manipulating the air door, there should be a detectable difference in the strength of the air draft from the low setting to the high setting.

Often, the first thing that folks do when their refrigerator starts to feel warm is turn both controls on the coldest settings. **This is exactly the WORST thing to do.** Turning the cold control to the coldest setting *will* keep the compressor running longer and make lots of cold air. But turning the air door to the coldest setting *closes* the airway to the food section. Lots of cold air is made, but it all stays in the freezer section, and the food section actually gets *warmer*.

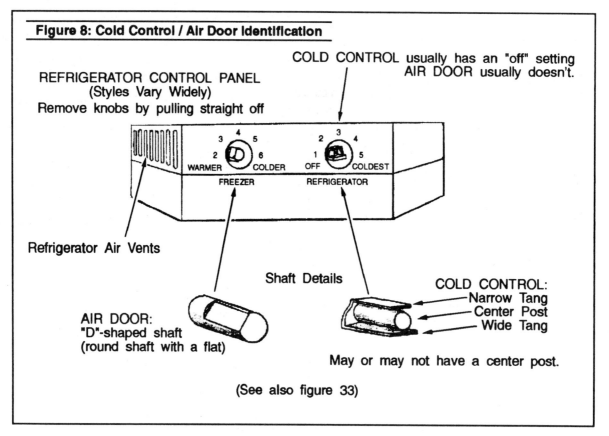

(See also figure 33)

## OTHER CONTROLS: BUTTER CONDITIONER, ENERGY SAVER, MEAT OR VEGETABLE COMPARTMENT CONTROLS

*EFFICIENCY OR ENERGY SAVER SWITCHES:* If you live in a warm, humid environment, you may have a problem with condensation forming on the *outside* of your refrigerator. This is known as "sweating." So-called "Energy Saver" switches control small, low-wattage "mullion" heaters in the side and door panels that prevent the outside of the refrigerator from getting cool enough for sweating to occur. In the "economy" position, the heaters are *off*.

*MEAT or VEGETABLE (CRISPER) COMPARTMENT CONTROLS:* These may be small mullion heaters or they may be separate air doors that control the airflow within these compartments. The idea is to keep the compartment at a different temperature from the rest of the food compartment; a more optimum temperature for the particular food that you're keeping in these compartments.

*BUTTER CONDITIONERS:* Again, small heaters that keep the butter compartment at a different temperature than the rest of the food compartment.

## 4-2. CONDENSER AND CONDENSER FAN

Next, check your condenser and condenser fan.

The locations of the most common types of condensers are shown in Figures 9A & 9B. Any type condenser mount may be used on bottom-freezer, top-freezer or side-by-side units.

**FIGURE 9A**: A back-mounted condenser has no condenser fan. Air flows over it by convection"the warm air rises and is replaced by cooler air from below. Some of these condensers are covered by a metal plate.

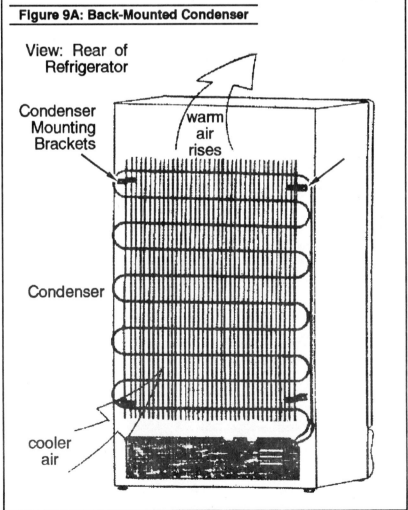

**FIGURE 9B**: Bottom-mounted condensers come in many configurations. Most look like a radiator or grille beneath the fridge, behind the kickplate. They are accessible for cleaning through the bottom front of the refrigerator.

**FIGURE 10**: Another fairly common type of bottom-mount condenser is wrapped in a metal plate and is accessible through the bottom back panel of the refrigerator.

The condenser *fan* may be mounted in a number of different ways. Usually it is accessed by removing the bottom back panel. Figures 10 & 11 show the most common arrangement for the condenser fan.

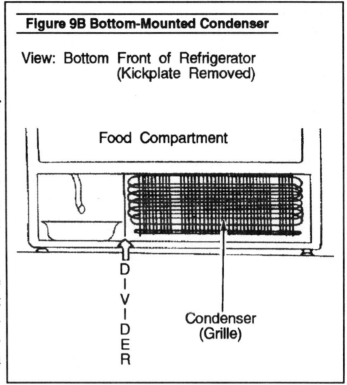

Figure 9B Bottom-Mounted Condenser

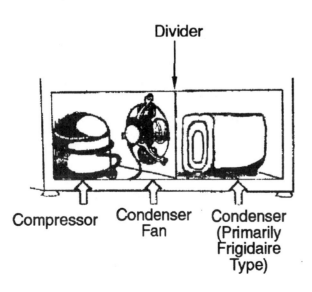

Figure 10: Compressor (Most Models)
Condenser Fan (Most Bottom-Condenser Models)
Condenser (Frigidaire-Type)

## DIAGNOSIS AND REPAIR

If you have a back-condenser refrigerator, make sure that nothing has fallen behind your fridge that might block the airflow.

If you have a bottom condenser, remove the baseplate (kickplate) from the bottom front of the refrigerator and look beneath it with a powerful flashlight. If you have kids or dogs or if your clothes dryer is installed nearby, you are a prime candidate to have a blocked condenser. Quite a bit of dust is normal; still, enough may be impacted to block the airflow completely. Feel for a steady flow of warm air from the drain pan side; it should be obvious (see Figure 11). Clean your condenser with a condenser brush. (*NOTE: Some condensers must be accessed through the back panel.*)

## COMPRESSOR IS RUNNING BUT REFRIGERATOR IS NOT COLD

*CAUTION: When cleaning your condenser, you want to do it thoroughly, but don't be too vigorous about it. You don't want to knock loose any wiring beneath the fridge. Also, you may hit the condenser fan (with a grinding thud) with the brush. Don't worry too much about it; you probably won't hurt the fan or motor, though it's not something you want to actively pursue.*

If the airflow improves dramatically, that may have been your only problem.

After you clean your condenser, pull the lower back panel off the fridge. Make sure that the condenser fan is running and not blocked by any loose insulation or other objects.

**Figure 11: Bottom Condenser Cut-Away View: General Arrangement, Airflow, and Cleaning**

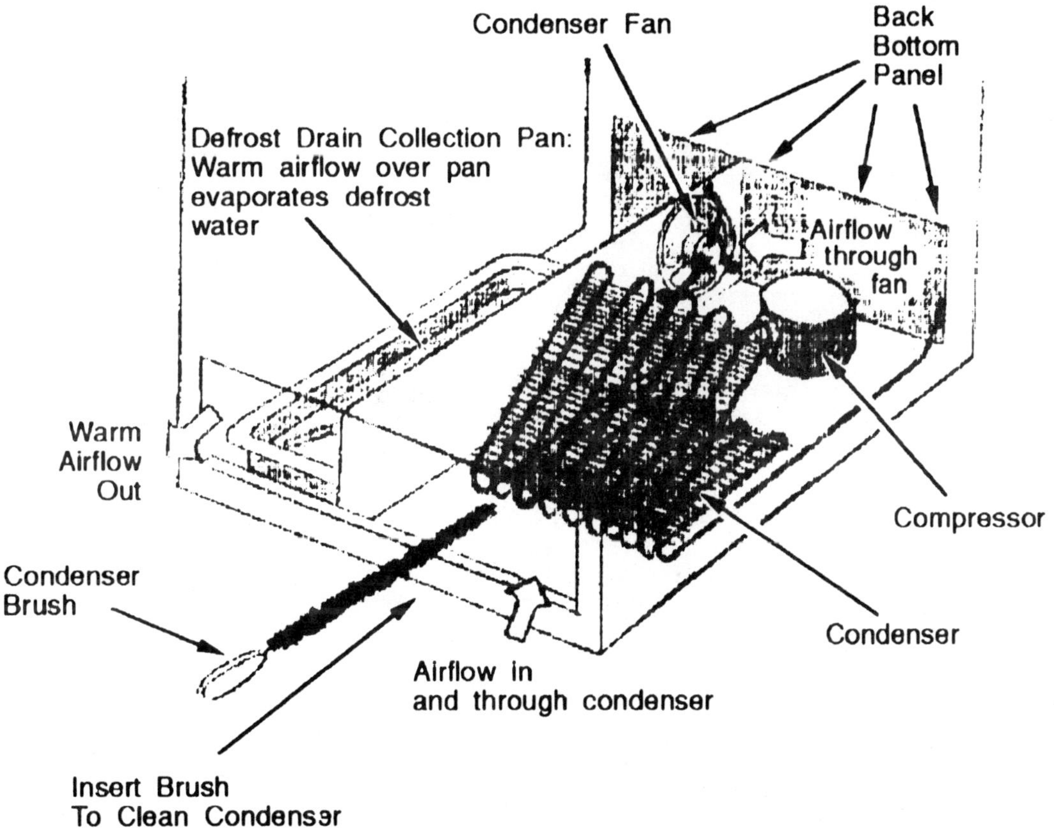

- Condenser Fan
- Back Bottom Panel
- Defrost Drain Collection Pan: Warm airflow over pan evaporates defrost water
- Airflow through fan
- Warm Airflow Out
- Compressor
- Condenser Brush
- Condenser
- Airflow in and through condenser
- Insert Brush To Clean Condenser

MAKE SURE THAT YOU REPLACE THE BACK BOTTOM PANEL. If it is missing, fashion one out of a piece out of corrugated cardboard and screw it on using the existing screwholes. It has the very important job of directing airflow beneath the fridge, assuring that the condenser fan is drawing air over the condenser and not just sucking air in through the back of the fridge. (Figure 12)

If the condenser fan is stopped and there is nothing blocking it, replace the fan motor. They are sealed units and cannot be rebuilt.

Replacing the condenser fan motor can be dirty and difficult. There are two types of mounts most commonly used; (Figure 13) bracket

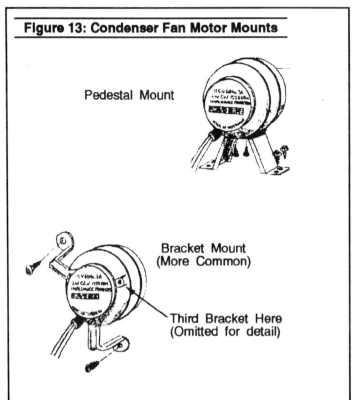

Figure 13: Condenser Fan Motor Mounts

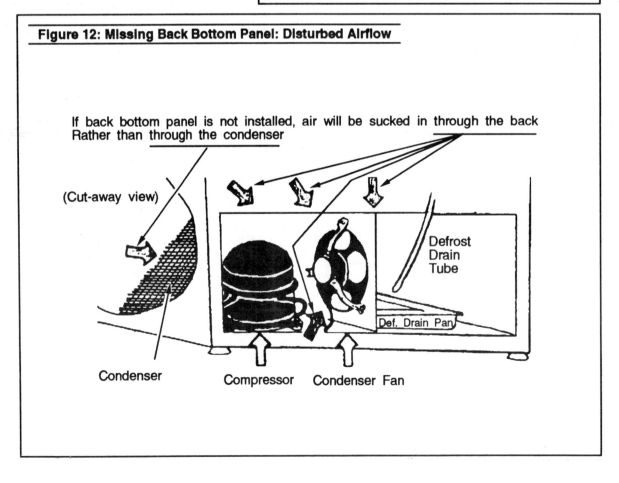

Figure 12: Missing Back Bottom Panel: Disturbed Airflow

mount (the most common) and pedestal mount. It is usually easiest to remove the mounting brackets or pedestal from their place in the refrigerator, with the fan motor attached. When installing a new motor with bracket mounts, it may be easier to install the brackets loosely on the motor until you can locate the mounting screws in their holes. The wires can be cut and reconnected with crimp-on butt connectors or wire nuts and electrical tape.

Even if you find a dirty condenser or stopped condenser fan, it's a good idea to go through the rest of the tests in this chapter to be certain that you've solved your problem.

## 4-3. EVAPORATOR FAN

On some models, the evaporator fan shuts off via a door switch when you open the refrigerator door. Thus, when troubleshooting the evaporator fan, you must depress the door switch(es).

Open your freezer door, depress all door switches and listen for the evaporator fan. If you do not hear it running, there's a problem. It might be ice-blocked, or it might have worn out and stopped. The door switch that operates it might be defective. If you *do* hear the evaporator fan running but you do not feel a strong blast of freezing air from the freezer vents, then you probably have a frost problem; see section 4-4.

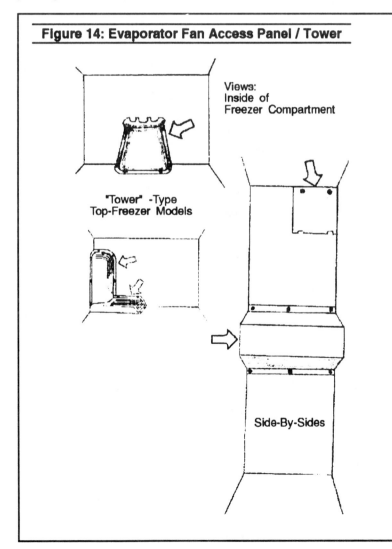

**Figure 14: Evaporator Fan Access Panel / Tower**

**IF THE EVAPORATOR FAN IS NOT RUNNING,** You *may* need to pull off the entire evaporator panel as described in section 4-4 to access the fan. Look first for a separate access panel or a tower within the freezer that houses the fan (Figure 14). Check for anything that may be blocking the fan, including ice from a backed-up defrost drain or a frost problem. If something is blocking the fan, clear it out of the way. If the blockage is due to frost or ice, you must investigate the source and solve the problem.

Depress the door switch. If nothing is blocking the fan and it still does not run, check for voltage across the fan motor leads (with the door switch depressed, of course.)

If you have voltage across the fan motor leads, the fan motor is bad. Replace it.

If you *don't* have voltage to the fan motor, the door switch might be bad. Take power off the fridge and pry out the door switch. You might have to destroy it to get it out. Check for voltage to the switch. If the switch is bad, replace it. (Figure 15).

## SLOW-RUNNING EVAPORATOR FAN MOTOR

Sometimes the evap fan will run *slower* than it should. This can be difficult to diagnose. It *can* cause ice to build up in the internal ductwork.

If you hear a "whistling" or "warbling" noise emanating from the fan motor itself or from the inside of the food or freezer compartment, it is probably coming from the evaporator fan motor. The bearings are worn and loose or sticky. Replace the motor.

As I said, a slow evap fan can be very difficult to diagnose. Usually it is done by sound and by experience. The chances are, if it *sounds* slow or strange, it is malfunctioning. Try replacing it. They don't cost much.

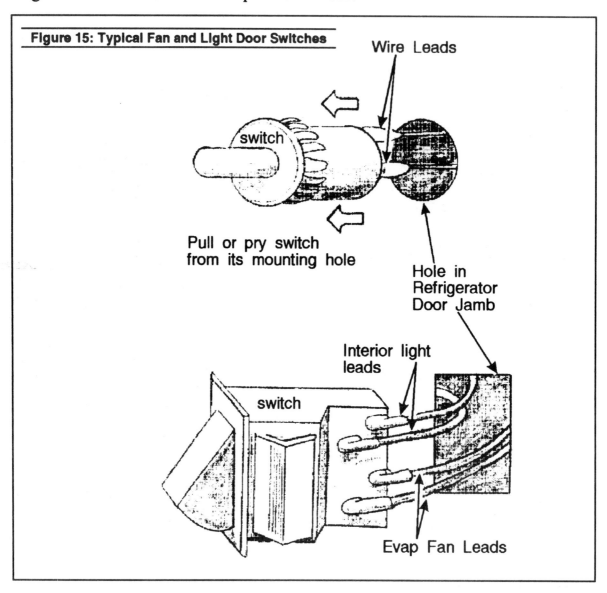

Figure 15: Typical Fan and Light Door Switches

## REPLACING THE EVAPORATOR FAN MOTOR

In replacing the fan motor, you must make sure that the rotation of the new fan motor is the same as the old one. The easiest way to do this is to look for the shading poles on the old fan motor (Figure 16). If they are on opposite corners from the ones on the new fan motor core, it is a simple enough task to reverse the new rotor in its core. Carefully remove the bearing cage screws and simply turn the rotor around so the shaft sticks out the other end of the motor.

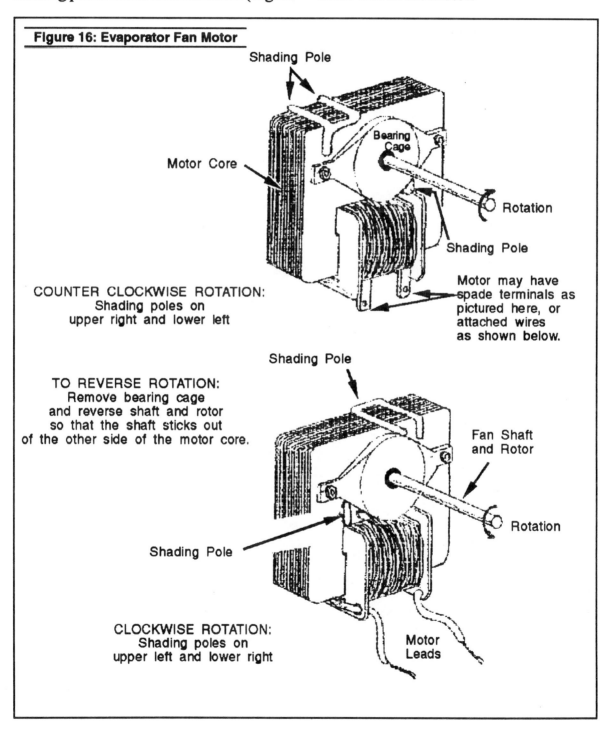

Figure 16: Evaporator Fan Motor

## 4-4. FROST PROBLEMS

Remove everything from your freezer, including all food and any shelves. Do not remove the icemaker (if installed.)

Look at and feel the panel covering the bottom or back of the freezer compartment. Is it thick with frost?

On top freezer models, are the holes in the top of the food compartment that lead to the evaporator choked with ice? Is there ice forming on, or lots of water on the ceiling of the food compartment?

If the answer is yes to any of these questions, there's probably a defrost problem.

If you suspect a defrost problem, first remove any icemaker that may be installed. You will see a removable panel covering the entire back or bottom of your freezer

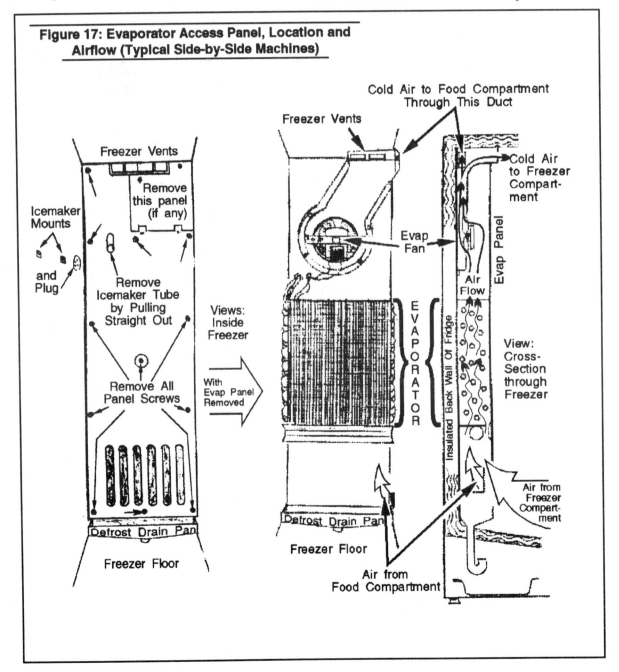

Figure 17: Evaporator Access Panel, Location and Airflow (Typical Side-by-Side Machines)

compartment. There may be 6 to 10 or more screws holding it on. (Figures 17, 18 & 19) In some units there is a light socket attached to the panel; you'll see this arrangement most often in side-by-sides. These can sometimes be quite difficult to disassemble. Make sure the power is off the refrigerator before disassembling any lighting circuit.

On some bottom-evap models, you may have to remove some of the plastic moulding around the door frame to access some of the evaporator panel screws. (Figure 19) Be extra careful; the plastic moulding can be brittle and break easily. The plastic will bend; just go slow. You may even try heating it a little with a blow dryer, to soften it.

The panel may be frozen to the evaporator; be careful you do not bend or break it. Sometimes it pays to take a few extra minutes and melt the ice a little bit first. This can usually be accomplished by placing a pan of very hot water in various places on the panel, or by blowing warm air on it with a blow-dryer. Do not melt *all* the ice just yet; only enough to get the panel off. You want most of it to remain there at this point so you can further diagnose the problem.

*NOTE: The terminating thermostat opens at a temperature of somewhere between 40 and 90 degrees F, depending on the design of your fridge. Most are between 50 and 70. It does not close again until well below 32. Ice is cold enough to keep it closed, but not to close it again if it opens. Therefore, when you are diagnosing a defrost problem, it's a good idea to try to avoid melting the ice encasing the terminating thermostat until you've made your diagnosis. If the thermostat opens before*

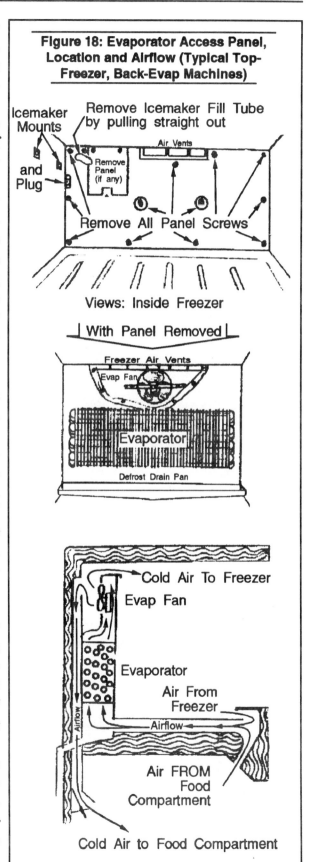

Figure 18: Evaporator Access Panel, Location and Airflow (Typical Top-Freezer, Back-Evap Machines)

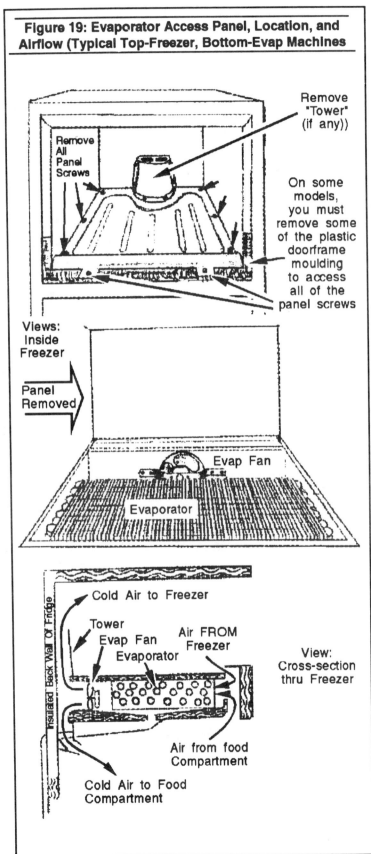

Figure 19: Evaporator Access Panel, Location, and Airflow (Typical Top-Freezer, Bottom-Evap Machines

you've had a chance to see if the heater works, you'll have to bypass it. On some models, this involves cutting, stripping and splicing wires. No big deal, but it's an extra step that's unnecessary if you're careful about melting ice in the first place.

Some of the styrofoam insulation panels may be waterlogged and may break when you remove them. It's okay, just keep them in one piece as much as possible and replace them as best you can when you're re-assembling.

**EVAPORATOR TYPES**

There are countless different arrangements for the evaporator and its fan and ducting, but almost all arrangements are relatively simple and easy to trace.

The evaporator looks like a group of looped aluminum tubes, usually with fins attached. The fins are sharp; be careful not to cut your hands on them.

Chill-plate evaporators look like a silver aluminum plate, sometimes in the form of a box. There are many styles, but most are variations of the three types pictured in Figure 20.

A *back-evaporator* model is one with the evaporator mounted vertically against the *inside back wall* of the freezer compartment. These may be bottom freezer models, side-by-sides (Figure 17) or top freezer models (Figure 18.)

A *bottom-evaporator* model is one with the evaporator mounted horizontally (flat) beneath a panel on the *bottom* of the freezer compartment (Figure 19). These are generally top-freezer models only.

**WHEN YOU GET THE PANEL OFF**, examine the *quality* of the ice that's built up on the evaporator. Is it frosted heavily enough to block the airflow, or is it just a thin white coating? Does it have a fluffy (snowy) white consistency, or is it solid and clear-ish or slightly milky whiteish?

Check the frost *pattern*. Is the evaporator frosted on one or two coils, and then clear on the rest? Or is it pretty evenly frosted? Or is it not frosted at all?

On back-evap models, examine the drain pan directly beneath the evaporator. Is it clear, or is it filled with solid ice?

Each of these symptoms indicates a different problem. If you have solid ice, see Chapter 6. If no coils are frosted, or just one or two, see "UNEVEN FROST PATTERNS," section 4-8. If you see a thin, even, white coating of ice on the evaporator, and no ice in the defrost drain pan, the defrost system is probably O.K.; go to section 4-9. If you have lots of white, snowy ice, keep reading.

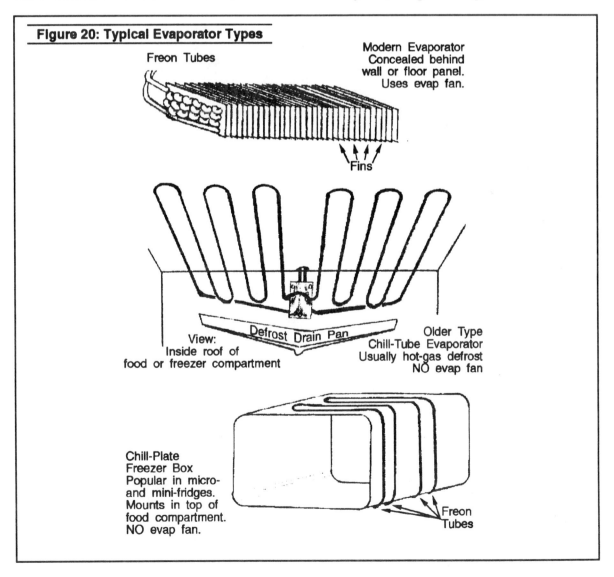

## 4-5. DEFROST SYSTEM

If the frost is snowy and white in appearance, you have a defrost problem. The three main components of the defrost system are the defrost timer, the defrost heater and the terminating thermostat.

### 4-5(a). DEFROST TIMER

The defrost timer can be a bit difficult to find. They come in many different styles. Often they are mounted under a cover plate or in a bracket that hides all but the advancement pinion. Figure 21 shows some different style timers and what the timer might look like installed; Figure 21A shows some typical mounting locations.

There is one **GLARING** exception. If you have a Whirlpool or Kenmore refrigerator with a flex-tray icemaker, the defrost timer is integrated into the icemaker. It is NOT a separate unit. This is true whether you are using the icemaker to make ice or not—it is running constantly to time your defrost cycles. Follow the instructions in section 4-6.

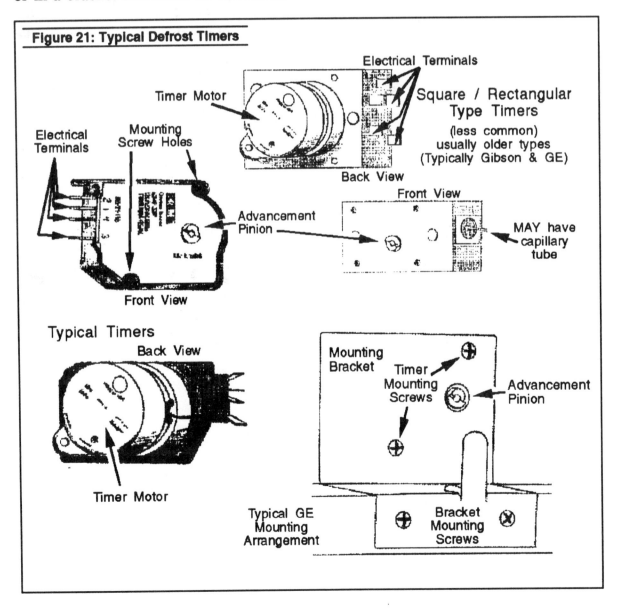

Figure 21: Typical Defrost Timers

# COMPRESSOR IS RUNNING BUT REFRIGERATOR IS NOT COLD

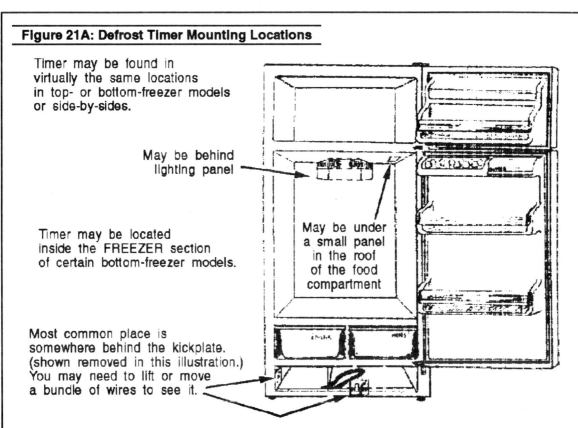

**Figure 21A: Defrost Timer Mounting Locations**

Timer may be found in virtually the same locations in top- or bottom-freezer models or side-by-sides.

May be behind lighting panel

May be under a small panel in the roof of the food compartment

Timer may be located inside the FREEZER section of certain bottom-freezer models.

Most common place is somewhere behind the kickplate. (shown removed in this illustration.) You may need to lift or move a bundle of wires to see it.

Front of Refrigerator

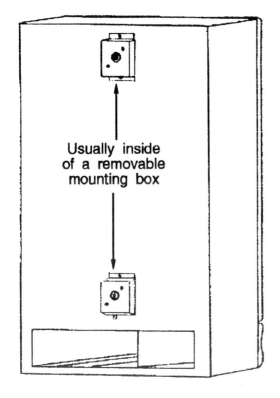

Usually inside of a removable mounting box

Rear of Refrigerator (less common)

## 4-5(b). DEFROST HEATER

Defrost heaters are always located in the evaporator compartment. See Figures 22 A, B, C, & D for arrangement and types. There are three different types most commonly used:

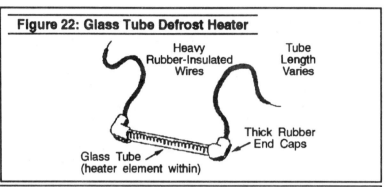

Figure 22: Glass Tube Defrost Heater

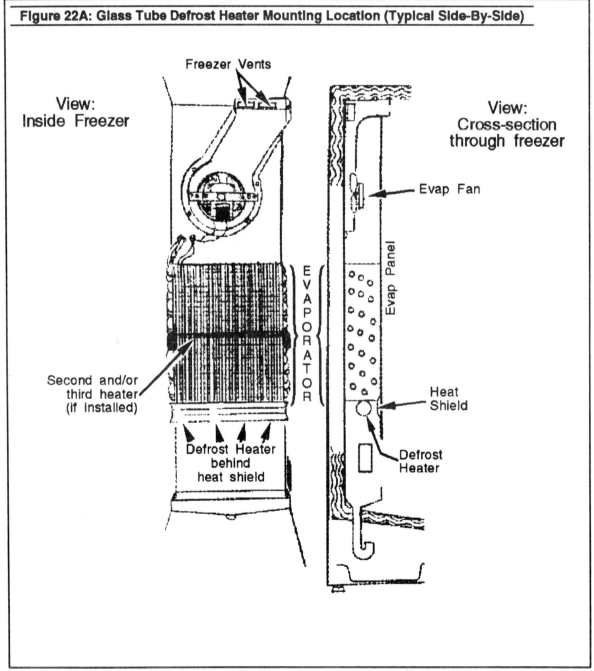

Figure 22A: Glass Tube Defrost Heater Mounting Location (Typical Side-By-Side)

**22, 22A, 22B) Glass-tube defrost heaters:** The heating element is encased in a glass tube mounted beneath the evaporator. Sometimes two or three small glass-tube-type heaters will be used instead of one big one; usually you'll see this arrangement in side-by-sides.

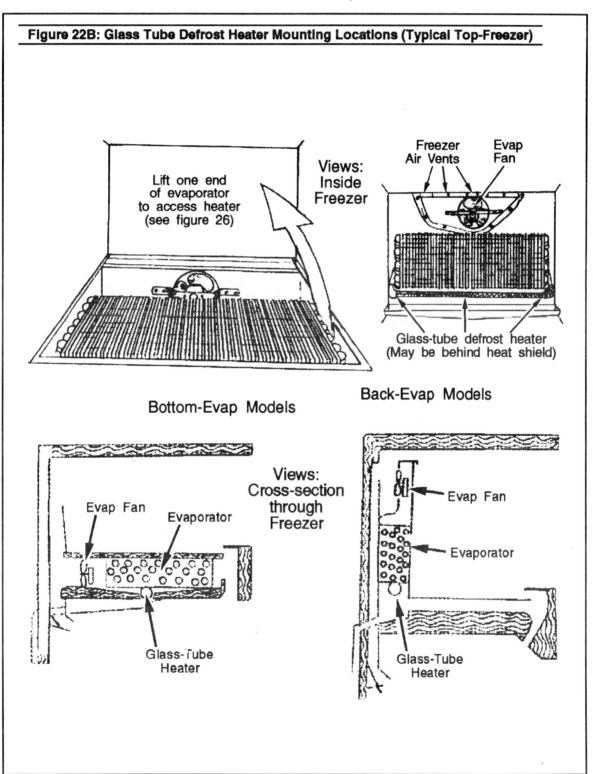

Figure 22B: Glass Tube Defrost Heater Mounting Locations (Typical Top-Freezer)

**22C) Aluminum tube heaters:** These heaters look just like the evaporator tubes and press into the evaporator fins. They are usually used on bottom-evap models. The easiest way to see the heater is to look for the heavy, rubber-coated wires leading to it; one on each end. Often there are clips holding the ends on to the evaporator coils; watch for these when you remove the heater.

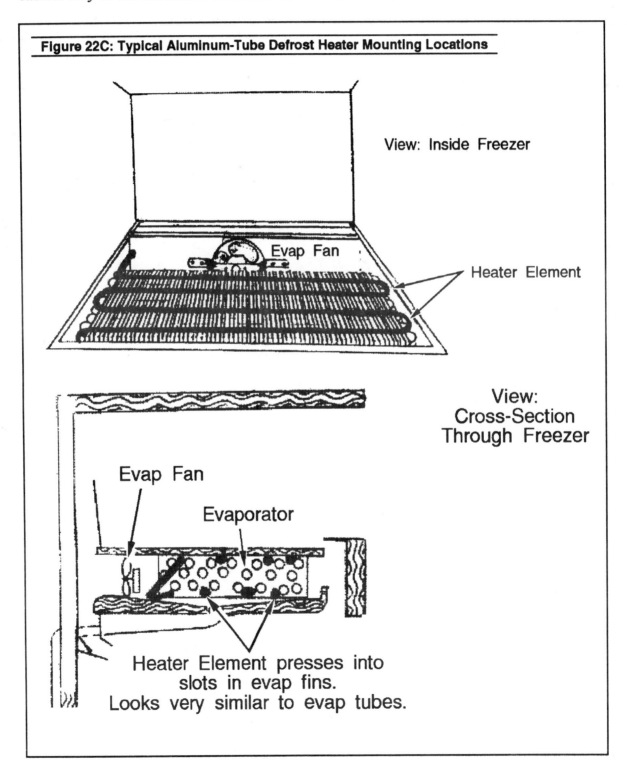

Figure 22C: Typical Aluminum-Tube Defrost Heater Mounting Locations

**22D) Bare element heaters:** Found most commonly on top-freezer back-evap models. The element has no protective tubing and generally wraps around beneath the evaporator in a large "U" shape.

You must exercise caution when handling these heaters to prevent burning yourself. They all run very hot; glass tube and bare element heaters even glow red while in operation.

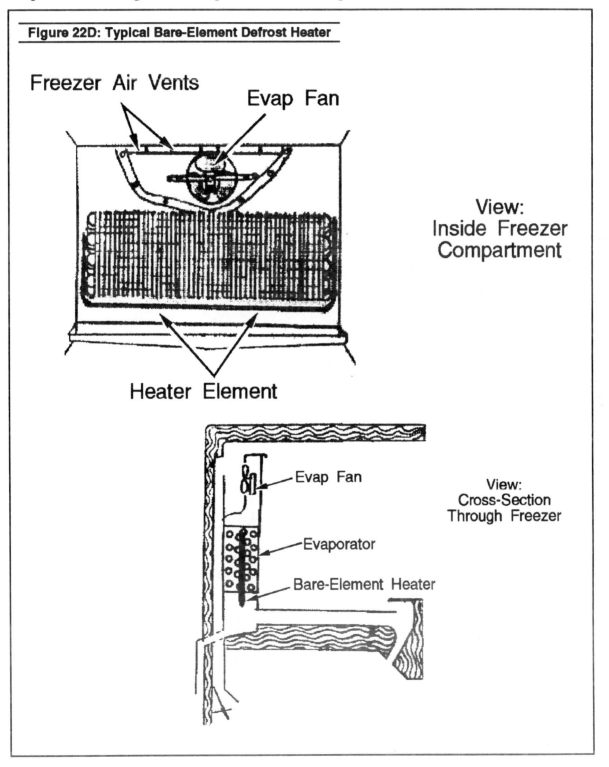

Figure 22D: Typical Bare-Element Defrost Heater

## 4-5(c). TERMINATING THERMOSTAT

A terminating thermostat will also be located somewhere in the evaporator compartment, usually to the evaporator itself (by a spring clip) or against the side or back wall of the compartment. It looks like a small cylindrical disc about 1" or so in diameter and about 3/4" to 1" thick (Figure 23.)

It is wired in series with the defrost heater; when it opens, the heater shuts off. One of the two heater wires will lead directly to it.

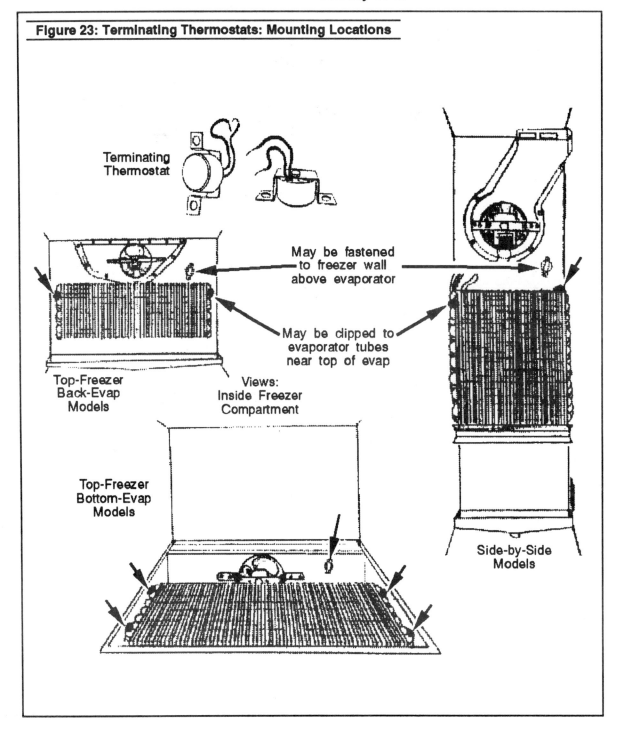

Figure 23: Terminating Thermostats: Mounting Locations

## 4-5(d). DEFROST SYSTEM DIAGNOSIS AND REPAIR

The first thing to do is to find your defrost timer. (See section 4-5(a).

**ADVANCING THE DEFROST TIMER**

When you find your timer, put a screwdriver in the advancement pinion and advance it (clockwise *only*, or you will break it). Sometimes it takes a pretty firm twist to advance it. You will feel it clicking. At some spot in the cycle, you will hear and feel a loud click; after you advance it 10-20 more degrees or so, you will feel and hear another loud click. (Figure 24)

Between the two loud clicks is the defrost part of the cycle. The rest of the timer's rotation is the "run" cycle. If your compressor is running when you advance the defrost timer, it will stop running when you hit the defrost portion of the cycle.

Advance the timer all the way around to the beginning of the defrost cycle again (generally one-half or one full turn) and leave it as early in the defrost cycle as possible.

If you have a hot-gas defroster, go now to section 4-7.

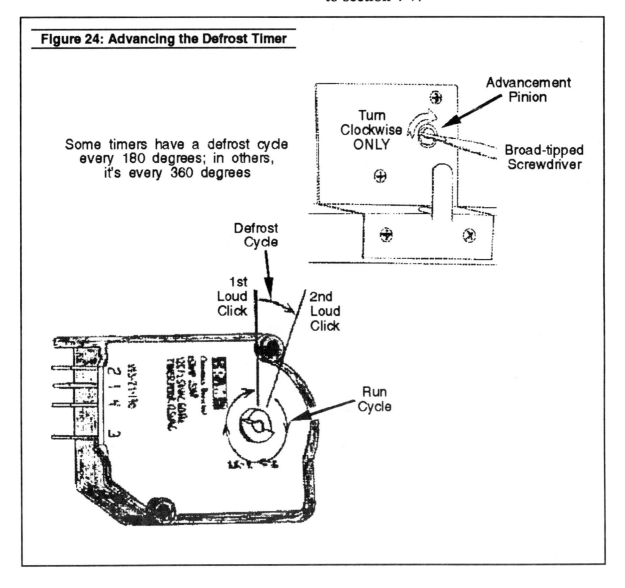

Figure 24: Advancing the Defrost Timer

## DEFROST TIMER

Look and listen to the evaporator. Within ten minutes (usually much less) you should be able to see a red glow from the defrost heater(s), which is (are) mounted beneath the evaporator. If you have an aluminum-tube heater as described in section 4-5(b), it will not glow red, but you *will* see ice melting away from its coils. Be careful; all defrost heaters run hot enough to burn you. You will probably also hear popping and sizzling; this is defrost water hitting the heater and boiling off.

If you *do* see or hear any of these indications, the problem is the defrost timer; it is not advancing. It can get old, worn and coked up with dust, and it may develop hard spots in the bearings.

Unplug your fridge and replace the timer. If the timer is connected by a terminal block, it probably plugs in directly. If you have separate wires to the timer terminals, carefully record which wire came off which terminal, by color or by terminal number, or both. Draw a picture, if you have to. Make sure that the new timer is wired correctly; there should be instructions with the new timer.

**Figure 25: Whirlpool / Kenmore Defrost Timer Motor Wire**

Loose wire goes either on terminal 1 or terminal 2. To determine which:

1) Check the old timer
2) Check a wiring diagram
3) Ask your parts dealer

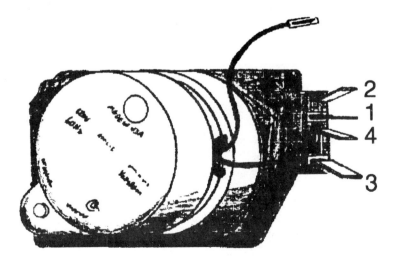

*CAUTION: If you have a Whirlpool or Kenmore timer with a separate wire coming from the timer motor, (Figure 25) it is important to get that wire connected to the proper terminal. If the wire is visible on the old timer, connect it to the same terminal. If you cannot tell for sure, get the information for your model fridge from your parts man. If the timer is wired incorrectly, the fridge will frost up again.*

Thoroughly melt all the ice in the evaporator compartment and in the air ducts leading to and from the compartment, and re-assemble your fridge.

## DEFROST HEATER AND TERMINATING THERMOSTAT

If you do not hear or see indications that the defrost heater is working, then it is necessary to investigate a little further. You could be looking at one of several different problems. The heater element *may* be burnt. The heater *may* be so icebound that it would take *hours* for the heater to melt enough ice for you to see the heater begin to work. The terminating thermostat might be open.

Whether you have an ammeter or not, if you think the defrost heater is not working, test it for continuity.

If you don't have an ammeter, thoroughly melt all the ice in the evaporator compartment and find the proper power leads for your heater. If they are not connected to a terminal block, you will need to cut the leads to test for continuity. Make sure you're not testing continuity across the terminating thermostat too; it may be wide open above 40 or 50 degrees. If the heater shows good continuity, it is working fine. If you have more than one heater, test each, unless they are permanently wired in series. If they are permanently wired in series, test them as a set. If the heater is bad, replace it. If you have multiple heaters, replace them as a set. With glass-tube heaters, be careful that the glass is not cracked or broken and that you do not cut yourself. It is a good idea to replace the terminating thermostat with the heater(s)"it's cheap.

If you have an ammeter, try to determine if the heater is drawing any power before you melt any ice. If so, it will be drawing between about 2 and 5 amps. In trying to find the heater leads, be careful that you do not melt so much ice that the terminating thermostat opens. If you suspect that the terminating thermostat might be open, temporarily bypass the terminating thermostat with an alligator jumper as described below.

If you cannot find the heater leads, an alternative is to check the current in one lead of the main power cord. (See Figure 7) If the fridge is in the defrost cycle and the interior lights are off, then the only current draw will be the defrost heater.

If the heater(s) test out O.K., then the problem is your terminating thermostat. Double-check this diagnosis by jumping across (shorting) the terminating thermostat with your alligator jumpers. If the two thermostat leads are not on a terminal block, you will have to cut the leads to jump the thermostat. Start the defrost system as described earlier in this section. If the defrost heater now heats up, your terminating thermostat is definitely bad. Replace it.

In replacing heaters and/or terminating thermostats, you can use butt connectors, wirenuts, and electrical tape, or spade connectors if fitted. Remember that it's a wet environment.

## LIFTING THE EVAPORATOR

If you have a bottom-evap model fridge, replacing the heater will involve the delicate task of lifting the evaporator up to get to the heater. (See Figure 26)

The Freon tubes leading to the evaporator will both enter the evaporator at one end. If you break or puncture one of those tubes, you're looking at a potentially expensive sealed system repair.

Thaw out the evaporator as thoroughly as is humanly possible. The closer the tubes are to room temperature, the more malleable the metal will be. Do not heat the tubes. If you are changing an aluminum-tube defrost heater, remove any clips holding it to the evaporator and loosen the top heater coils from the evaporator fins. The evaporator fins are sharp; be careful you don't cut yourself. Remove the evaporator mounting screws (if there *are* any) and gently lift up the end of the evaporator opposite the tubes. Prop up the evaporator with a blunt instrument (I use my electrical pliers or a flashlight) and change the heater. While you're in there, make sure the drain is clear as described in Chapter 6. Do what you went in there to do, but as much as possible, avoid moving the evaporator around *too* much.

When you finish, gently lower the evaporator back into place. Always re-install any little chunks of styrofoam or duck seal that you may have removed from *beside* the evaporator; they keep the air flowing *through* the evaporator rather than *around* it.

Re-assemble the fridge.

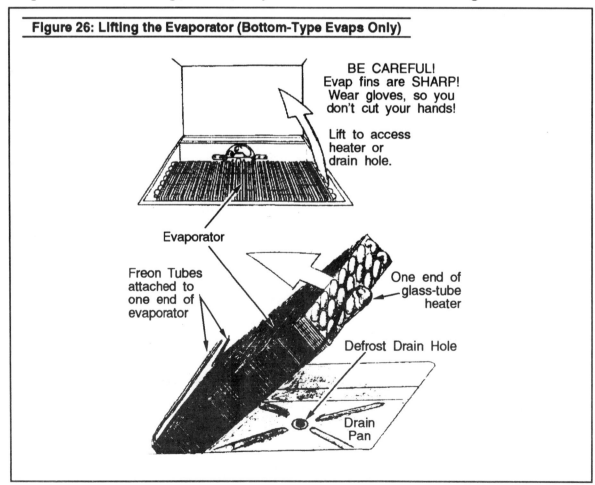

Figure 26: Lifting the Evaporator (Bottom-Type Evaps Only)

## 4-6. WHIRLPOOL/KENMORE FLEX-TRAY ICEMAKER DEFROST SYSTEM

You can recognize this type of defrost system by the shape of the cube it puts out (or *would* put out, if it was working). The hard-tray Whirlpool/Kenmore produces "half-moon" shaped cubes (see Figure 27.) The flex-tray produces "rounded rectangular" cubes. The hard tray is finished in a dark gray or black color and has rotating fingers that eject the cubes from the unit; the flex-tray has a white plastic, flexible tray that inverts and twists to eject, much the same as a manual ice cube tray would work. The hard-tray and separate defrost timer is by far the more common arrangement.

This defrost system has the same components described in the defrost system in section 4-5, except that the defrost timer is integrated into the icemaker. Whether it is being used to make ice or not, the icemaker motor runs whenever the compressor is running.

The defrost components will act the same, and you test them in the same manner as described in section 4-5, except for one item:

You cannot advance the icemaker like you can advance a defrost timer. However, the actual switch that controls the heater *is* accessible.

Remove the icemaker and the evaporator panel as described in section 4-4. *YOUR COMPRESSOR WILL STOP RUNNING WHEN YOU UNPLUG YOUR ICEMAKER. DO NOT BE ALARMED.*

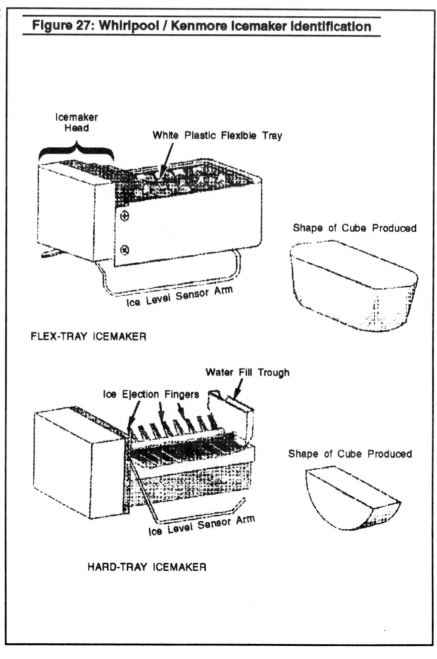

Figure 27: Whirlpool / Kenmore Icemaker Identification

Remove the ice tray from the icemaker. It is spring loaded and simply pushes away from the icemaker head and pops out. Take the plastic cover off the face of the icemaker and remove the three screws holding the metal faceplate to the icemaker head. (See Figure 28) Remove the drive cam, the large drive gear and the smaller timing gear. You will need to remove the leaf switch to get the drive cam off. Examine the gears for any stripped teeth. If you see any, replace the gears and drive pin as a set.

Temporarily remount the leaf switch to keep it from drifting around and touching things.

Inside your fridge, turn the cold control to its coldest setting. Plug the icemaker back into its electrical socket and observe the drive motor in the upper lefthand corner of the icemaker head. When the compressor is running, the motor will turn very slowly. If it doesn't, it is bad. Replace it.

Unplug your icemaker and look into the icemaker head. The defrost switch is the small, rectangular switch in the upper righthand corner of the icemaker head. Remove this switch from its mounts, but do not disconnect the wires to it. Using electrical tape, tape it out of the way so it does not touch any other metal object in the icemaker head. Plug in the icemaker again. *Do not touch any metal contact with your hands; you may shock yourself.*

Within a few minutes, you should start to see signs that the defrost heater is working as described in section 4-5(d).

**IF YOU SEE OR HEAR NO SIGNS OF THE DEFROST HEATER HEATING UP**, unplug your icemaker, remove the BLACK lead from the defrost switch and electrically test the switch for opening and closing. Using your resistance meter, you should see continuity (and no resistance) between the empty terminal (where the BLACK lead *was*) and the PINK terminal. You should see NO continuity between the empty (BLACK) and ORANGE terminal. When the switch toggle is depressed, continuity will be just the opposite: BLACK-ORANGE-CONTINUITY, BLACK-PINK-NO CONTINUITY. If the switch is not acting right, replace it. If the switch is okay, the problem is probably your defrost heater or terminating thermostat. Diagnose and repair as described in section 4-5(d).

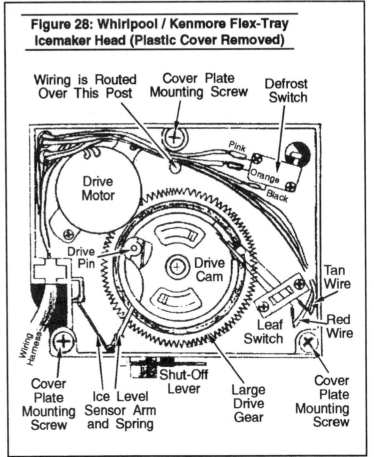

Figure 28: Whirlpool / Kenmore Flex-Tray Icemaker Head (Plastic Cover Removed)

## COMPRESSOR IS RUNNING BUT REFRIGERATOR IS NOT COLD

**IF YOUR DEFROST HEATER DID HEAT UP** when you dismounted the defrost switch, then you need to replace the gear sets in your icemaker. Get *both* sets of gears (timing gears *and* drive gears) from your appliance parts dealer. Alignment of the gears is critical; follow the instructions that come with the gear sets carefully. When you replace the gear sets, it is also a good idea to replace the defrost switch. You may or may not want to replace the drive motor. They *do* quit on occasion, but they *are* a bit more expensive. If you replace the motor, you will have to re-align the defrost timing gear mechanism.

### RE-ASSEMBLY

If you have not removed the defrost timing gear housing from the back of the icemaker head or the motor from the front of the head, you will not need to re-align the *defrost timing* gear mechanism. However, you *will* need to realign the *drive* gear mechanism.

Align the hole in the small drive gear with the alignment hole in the icemaker head and install the gear. Check alignment by inserting a 3/32" rod (a drill bit will do) into the holes to make sure they line up. See Figure 29. If they do not line up perfectly, momentarily plug the icemaker

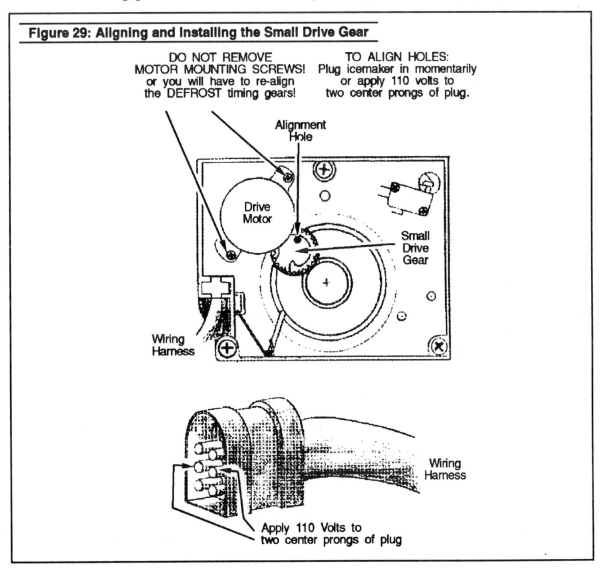

Figure 29: Aligning and Installing the Small Drive Gear

in or apply 110 volt power to the two center leads of the plug This will turn the drive motor slightly. Repeat the process until the holes align.

Install the large drive gear and align it on the same alignment hole. A second alignment hole is shown in Figure 30. The large drive gear must line up on *both* alignment holes *and* on the large drive cam hole in the center of the icemaker head.

Carefully holding the drive gear in its aligned position, install the drive cam (Figure 31.) Make sure the spring-loaded drive pin is in place in the cam and retained properly; the cam's spring-retainer should be in the pin's groove. Line the drive pin up on its hole on the drive gear. Lift the spring-loaded shut-off arm (ice level sensor) as you install the cam and let it rest in the cam hollow. Be sure that the ice level sensor arm loading spring is in the right place. Install the leaf switch. Sometimes the stuff in this paragraph takes three hands and your belly, but be persistent. You'll get it together.

Make sure the wiring for the leaf switch and the defrost switch is routed *over* the post above the drive gear. Carefully install the metal cover plate, making sure the end of the wire shut-off arm (ice level sensor) is in its pivot hole in the metal cover plate. Install your three screws. The drive pin will pop up through the metal cover plate.

Install the ice tray into the ice maker, and re-assemble your fridge.

The icemaker is now aligned at the beginning of an ejection cycle. When you re-install it, the ice tray will slowly turn 1 full turn. If the icemaker is being used, the tray will then fill with water. Make sure the icemaker is turned on (ice level sensor arm is down) or it won't make ice.

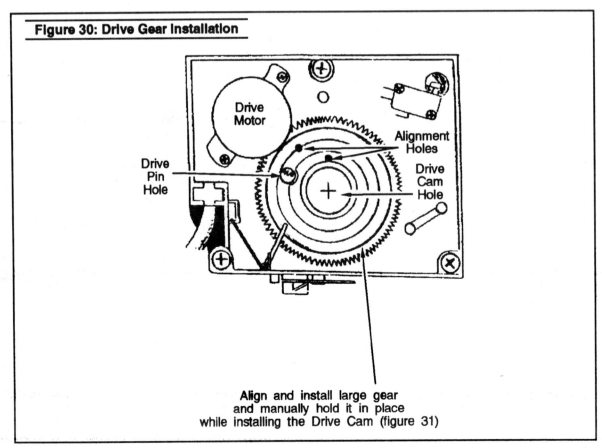

Figure 30: Drive Gear Installation

Align and install large gear and manually hold it in place while installing the Drive Cam (figure 31)

# COMPRESSOR IS RUNNING BUT REFRIGERATOR IS NOT COLD

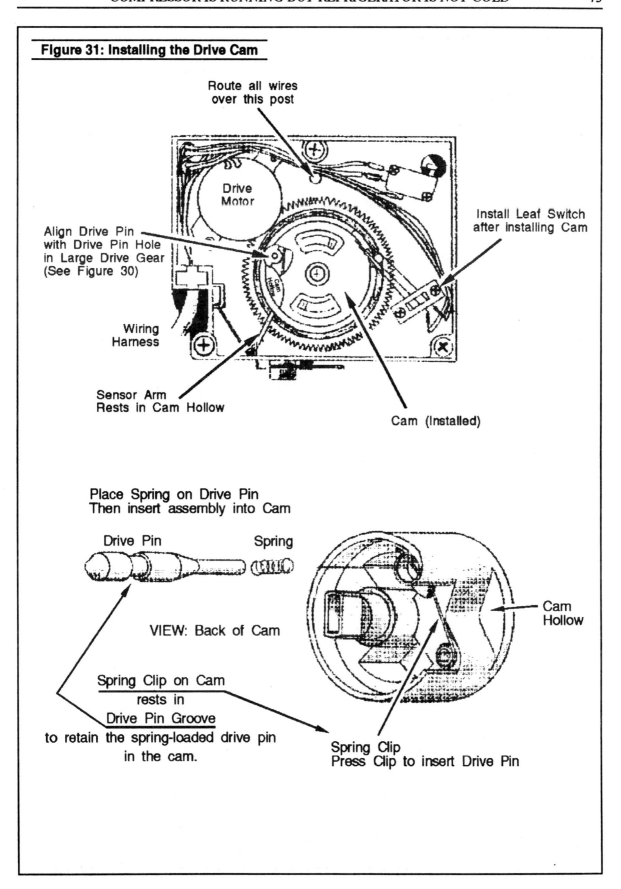

**Figure 31: Installing the Drive Cam**

## 4-7. HOT GAS DEFROST PROBLEMS

If you have a refrigerator with a hot gas defrost system, the defrost mechanism is somewhat different from those described in previous sections of this book. If you suspect that you might have a hot-gas defroster but you are not sure, ask your appliance parts dealer. Most of these refrigerators were built before 1970, but not all.

The main difference in a hot gas system is that there is no electrical heater or separate terminating thermostat. The defrost cycle is controlled by a defrost timer similar to the one you'll find in electric defrost systems, but the timer controls a *solenoid valve* instead of a *heater*. When this valve opens, it diverts the hot Freon gas coming out of the compressor. It then will flow through the *evaporator* instead of the *condenser*. The hot Freon gas flowing through the evaporator melts the frost from it. In order to supply the hot Freon gas needed to melt the frost, the compressor keeps running throughout the defrost cycle.

Another interesting feature of most of these refrigerators is that the defrost terminating thermostat was built into the defrost timer. These units have a temperature sensing bulb, similar to that found on the cold control (see section 4-9 and Figures 21 and 33) as a part of the timer. The sensing bulb is led to, and rests against, the evaporator. Its function is to sense the temperature of the evaporator so the defrost mechanism knows when to shut off.

Troubleshooting a defrost problem in this system involves two steps. First, find your defrost timer as described in section 4-5(a). Advance it into the defrost cycle as described in section 4-5(d). The compressor will *not* stop running. Wait and watch your evaporator for 10-15 minutes.

If the frost starts to melt, then your defrost timer has gone bad. Thoroughly melt the rest of the frost from your evaporator and replace the defrost timer.

If the frost does *not* start to melt, then your defrost solenoid is probably bad. Fortunately, the defrost solenoid is usually designed so the electrical coil can be replaced without cutting into the sealed system. You will find the coil behind the back bottom panel of the refrigerator. Trace the Freon tubing until you find electrical wires joining the tubing at a certain point. This will be the solenoid valve. (Figure 32). Replace the coil (solenoid) on the valve. Re-assemble your fridge.

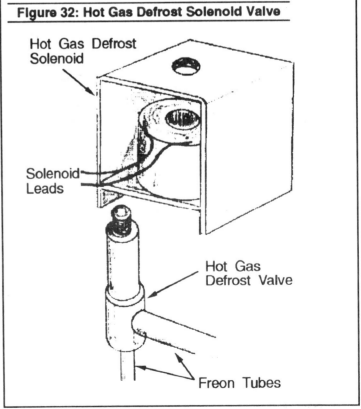

Figure 32: Hot Gas Defrost Solenoid Valve

## 4-8. UNEVEN FROST PATTERNS, OR NO FROST AT ALL

The evaporator should be bitterly cold to the touch. In fact, when you touch it, your finger will often *stick* to it. If the evaporator is either slightly cool or not cold at all, and your compressor is running *constantly* (not short-cycling; see section 4-9) you have a more serious problem. The same diagnosis applies if just the first coil or two in the evaporator is (are) frosted and the rest are relatively free of ice or perhaps even lukewarm.

What's happening is that the Freon is not getting compressed enough in the compressor. This could be due to two causes: either the amount of Freon in the system is low, or the compressor is worn out. It's time to call a technician out to your home, if you feel your fridge is worth saving. It *may* only require recharging the Freon system, which, at the time of this writing, should cost you about 50 dollars or less. I have only seen one exception to this diagnosis, and this is described in section 7-2.

Don't let the age of the refrigerator affect your diagnosis. Recently, one of the largest appliance companies put out a series of refrigerators with compressors that were either poorly designed or poorly constructed; I never did find out which. These were their giant, 20 to 25 cubic-foot flagship models, with techno-marvelous gadgets like digital self-diagnosis and ice and water in the door, and they were built with compressors that wore out within 2 years. Fortunately, the biggest and best companies warrant their refrigerators for five years or more, so these refrigerators were still covered under warranty. In my opinion, there *is* a real advantage to buying brand-name appliances.

## 4-9. COLD CONTROL

If your refrigerator is cold but not as cold as usual, and you cannot trace it to any of the other problems in this chapter, your cold control may be defective. To test its cut-in and cut-out temperatures, you can *try* putting the capillary bulb in ice water and measuring the temperature with a thermometer, but it's a wet, messy, job and it's difficult to control the temperatures. It's easier to just try replacing it and see if the fridge starts acting properly.

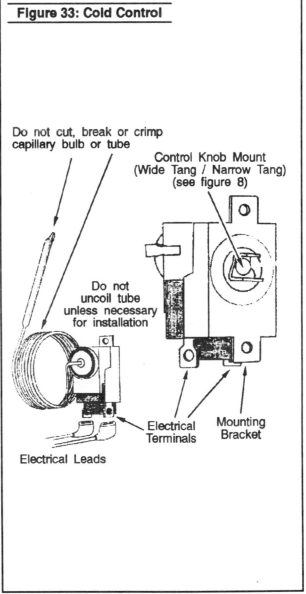

Figure 33: Cold Control

If you hear your compressor "short-cycling" (starting and stopping at short intervals) try jumping across the two leads of the cold control with an alligator jumper. If there is a green *third* lead, ignore it for this test; it is the ground wire. If the fridge starts running constantly, the cold control is bad. Replace it.

To test or change the cold control, first find it as described in section 4-1. Pull the knob off it and remove any plastic cover plate or housing from it. (Figure 33)

You will see two wires leading to it. There will also be a thick, stiff **CAPILLARY TUBE** attached. The capillary tube is the liquid-filled temperature-sensing element of the cold control, and operates in the same manner as a thermometer bulb; in fact, the end of the capillary tube may have a bulb. The tube and bulb *may* be coiled right next to the cold control, or they *may* be led away to another part of the compartment.

If you are just *testing* (electrically) the cold control, you can jumper directly from one wire lead to the other.

If you are *replacing* the cold control, it will be necessary to trace where the capillary tube goes, and remove the whole tube *with* the cold control. The new tube is replaced directly. Be careful not to kink the new tube (bend it too sharply) when installing it.

# Chapter 5 Overview
## Step-by-Step

Complaint: Warm refrigerator, or not as cold as usual
Chapter Qualifier: Compressor is not running

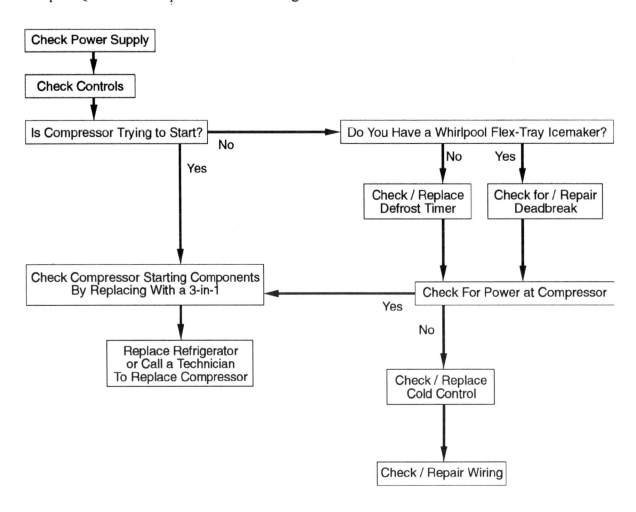

# Chapter 5

# REFRIGERATOR IS NOT COLD AND COMPRESSOR IS NOT RUNNING

## 5-1. POWER

If your refrigerator is not cold (or not as cold as usual) and you have determined that the compressor is *not* running (see the first page of Chapter 4,) first check that the fridge has power. If you have interior lights in the fridge, you have power. If you *don't* have interior lights, check your house breaker. Also check your wall outlet by unplugging the fridge and plugging in a portable appliance such as a blow-dryer or electric shaver.

## 5-2. CONTROLS

When you have established that power is getting to the fridge, check your controls. They sometimes have a way of getting magically turned off, especially in households with kids. (See section 4-1)

## 5-3. DIAGNOSIS AND REPAIR

Listen carefully to the fridge for a two or three minutes. If you hear a "CLICK-BUZZZZZZ-CLICK" (with a buzz of between about 5 and 30 seconds) then you *are* getting power to the compressor. See section 5-3(e).

## 5-3(a). DEFROST TIMER

If you hear nothing at all, set the cold control on the coldest setting and check the defrost timer. If it is stuck in the defrost mode, the compressor will not run. The terminating thermostat will open and stop the heater at a certain temperature, so the fridge will *not* heat up. You will hear, see and feel nothing but a fridge that's not cold. It's as if it wasn't even plugged in, except that the interior lights will still be on. (See section 4-5(d) and Figure 24 for details on how to advance the timer). Place a screwdriver in the advancement pinion and advance the timer manually (remember, clockwise *only*) about 1/4 to 3/8 of a turn. Does the compressor start? If it does, replace the defrost timer. If this does not start the compressor, make sure you leave the timer in the "run" mode for the rest of your diagnosis. Go to section 5-3(b).

In a Whirlpool flex-tray icemaker as described in section 4-6, the defrost timer is integrated into the icemaker. The defrost switch might fail into a "deadbreak" position in which nothing runs, similar to that described above. To test for deadbreak, first unplug and remove the icemaker from the freezer compartment. Look at the timing gear housing on the *back* of the icemaker head. You will see a white circle, about 5/8" in diameter (Figure 34). Examine the circle closely. You will see an arrow molded into the circle. If this arrow lines up with the arrow molded into the black timing gear housing, then you *might* have a deadbreak.

Test for continuity between the BLACK and PINK terminals and the BLACK and ORANGE terminals of the plug (see Figure 34). One or the other should show continuity. If neither does, you have a deadbreak. Replace the defrost switch and gear sets as described in section 4-6.

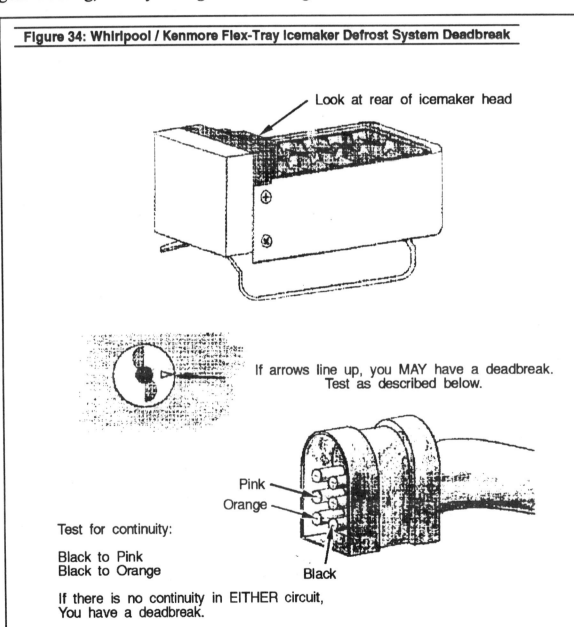

Figure 34: Whirlpool / Kenmore Flex-Tray Icemaker Defrost System Deadbreak

Look at rear of icemaker head

If arrows line up, you MAY have a deadbreak. Test as described below.

Test for continuity:

Black to Pink
Black to Orange

If there is no continuity in EITHER circuit, You have a deadbreak.

### 5-3(b). TESTING THE COMPRESSOR FOR POWER

If you *still* hear nothing at all, pull off the lower back panel of the fridge, remove the compressor relay cover. This is the square-ish plastic or bakelite box attached to the side of the compressor. Test for voltage at the two compressor leads. (Figure 35)

If you have voltage to the compressor but it is not starting, see section 5-3(e).

### 5-3(c). COLD CONTROL

If you don't have voltage to the compressor, use your alligator jumpers to connect the two wires of the cold control. (See section 4-9) If your compressor starts, the cold control is bad. Replace it.

**Figure 35: Testing for Voltage at the Compressor**

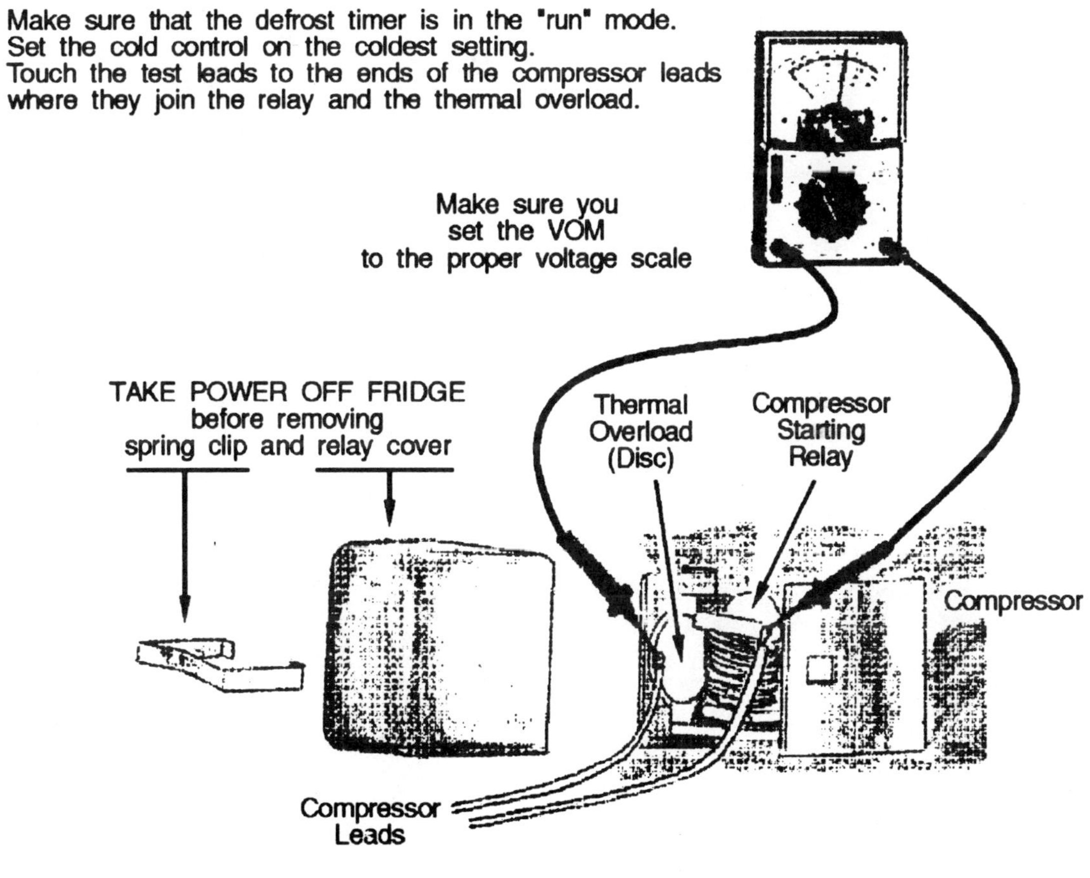

Make sure that the defrost timer is in the "run" mode.
Set the cold control on the coldest setting.
Touch the test leads to the ends of the compressor leads where they join the relay and the thermal overload.

Make sure you set the VOM to the proper voltage scale

TAKE POWER OFF FRIDGE before removing spring clip and relay cover

Thermal Overload (Disc)

Compressor Starting Relay

Compressor

Compressor Leads

## 5-3(d). WIRING AND ELECTRICAL

If the test in section 5-3(c) doesn't start the compressor, you're going to have to get a wiring diagram (there *may* be one pasted to the back of the refrigerator) and start tracing wires with your VOM to figure out where you're losing power. Check for dead mice beneath your fridge-- sometimes they get under there and start chewing wires. If you're thoroughly intimidated by electricity, then call a technician.

## 5-3(e). COMPRESSOR STARTING / ELECTRICAL COMPONENTS

If you *do* hear a sound, it will be something like: "CLICK-BUZZZZZZZZZZ-CLICK." The buzz will be between about 5 and 30 seconds long, and it will repeat within a minute or two. What's happening is that your compressor is trying to start, but it can't, and the electrical overload is cycling on and off. It might be that one of your compressor starting components is bad, or that your compressor motor is wearing or worn out. You may be able to test each starting component and replace the bad one, but I've found that the quickest and easiest way to diagnose this problem is to replace all three with a solid state "3-in-1" unit.

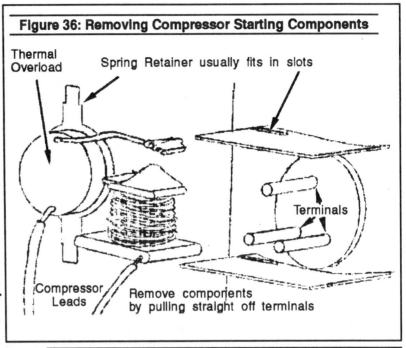

**Figure 36: Removing Compressor Starting Components**

Thermal Overload
Spring Retainer usually fits in slots
Terminals
Compressor Leads
Remove components by pulling straight off terminals

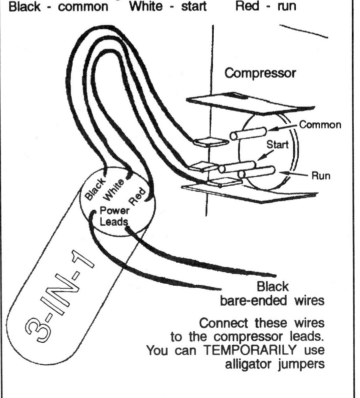

**Figure 37: Connecting a 3-in-1**

Press spade connectors onto compressor terminals. Follow the instructions that come with the 3-in-1. The USUAL wiring scheme is:
Black - common   White - start   Red - run

Compressor
Common
Start
Run
Black White Red Power Leads
3-IN-1
Black bare-ended wires
Connect these wires to the compressor leads. You can TEMPORARILY use alligator jumpers

First, remove the old starting components. (See Figure 36) Note carefully how they came off, in case you need to re-install them. Install the "3-in-1" using the instructions that come with it. (Figure 37). You can use your alligator jumpers to wire it in temporarily. Make sure the "3-in-1" you get is rated for the horsepower of your compressor.

*NOTE: Some of the latest models already have solid state starting components. If your refrigerator is this type, the "3-in-1" may fit onto your fridge directly, or it may not fit at all. If your case is the latter, you will need to get the original equipment replacement relay assembly for your fridge.*

If your compressor doesn't start with the "3-in-1," the compressor's dead. You will hear the overload in the "3-in-1" cycling just like the original overload did (CLICK-BUZZZZ-CLICK). It's time to call a tech for a compressor job, or to think about getting a new fridge.

If your refrigerator *does* start, unplug the refrigerator and wire your "3-in-1" in permanently. Use butt connectors, wire nuts and electrical tape. Make sure that none of the compressor terminals are touching each other or the metal housing of the compressor. Also make sure you cover the compressor terminals with a shield; usually you can use the old plastic relay cover and just lead the wires into it.

If the *cause* of your compressor's not starting was bad starting components, it will continue to run indefinitely.

If the cause of it not starting was that the compressor motor is getting worn out, the "3-in-1" will prolong the life of your compressor for somewhere between a few hours and a year or two.

You have no way of knowing which it was, or how long it will last, without some expensive tests that probably won't tell you much anyway. Count your blessings and start saving up for a new fridge (or a major repair), just in case.

# Chapter 6

# ICE OR WATER BUILDUP

If you have water build-up on the floor of the food compartment, or ice buildup on the floor of the freezer compartment, you are probably suffering from either a frost problem (see section 4-4) *or* from a clogged defrost drain. There *are* a few exceptions described in section 7-7.

## 6-1. DEFROST DRAIN SYSTEM

Directly beneath the evaporator will be a water collection pan with a drain hole. Leading from that drain hole to the drain pan beneath your refrigerator is a drain tube.

Side-by-side defrost drain tubes usually go straight down through the freezer floor to the drain pan. (Figure 38)

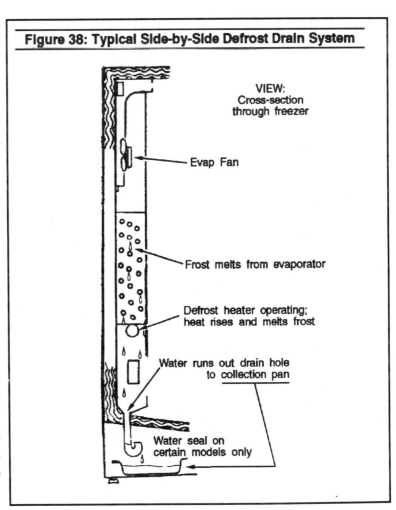

Figure 38: Typical Side-by-Side Defrost Drain System

Top-freezer models usually drain through a drain tube out the back of the refrigerator and down to the drain pan beneath the fridge (Figure 39). Some top-freezer models have a removable drain tube or trough inside the food compartment. (See Figure 39A)

Some refrigerators, especially older models, may have different defrost drain arrangements. Some top-freezer models drain the defrost water to the inside back wall of the food compartment. The water runs down the back wall to another drain hole which leads outside the fridge to a drain pan. (Figure 40).

**Figure 39: Typical Top-Freezer Defrost Drain System**

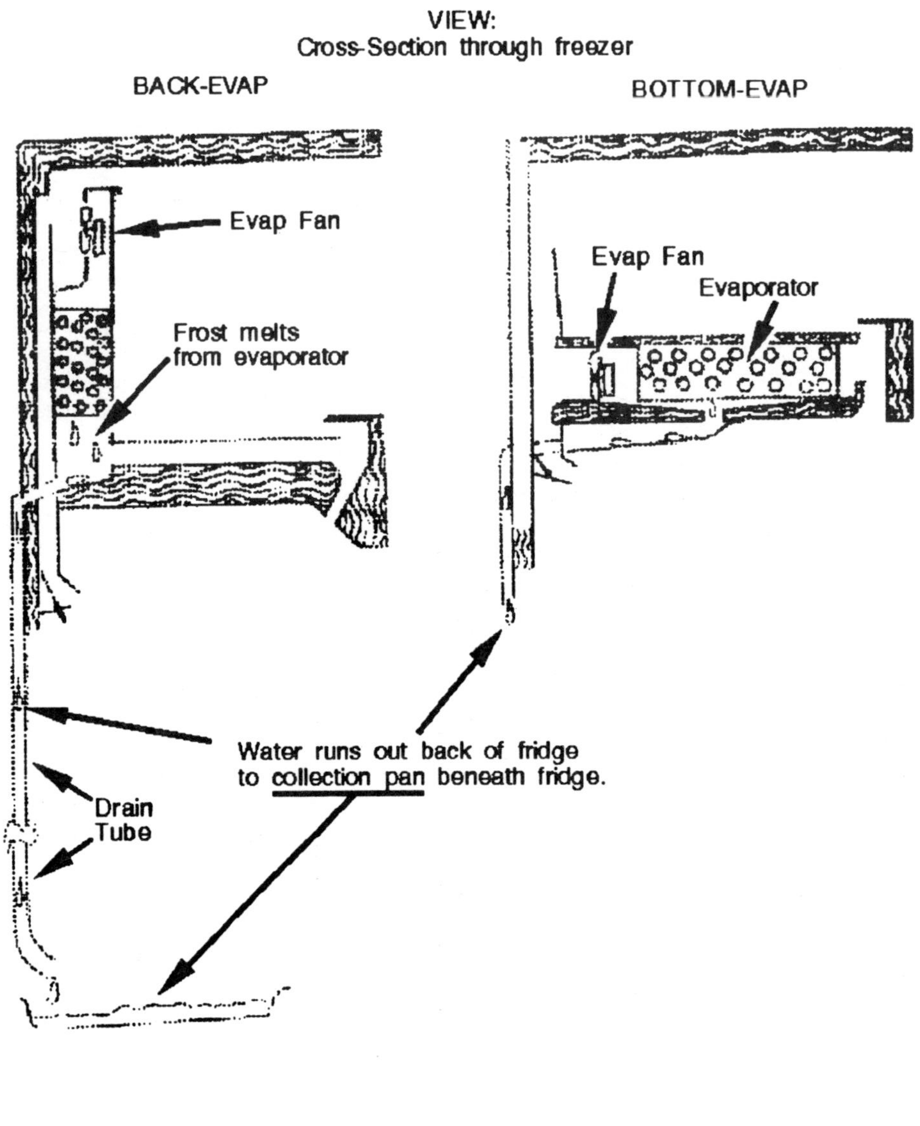

VIEW:
Cross-Section through freezer

BACK-EVAP

BOTTOM-EVAP

- Evap Fan
- Frost melts from evaporator
- Evap Fan
- Evaporator
- Water runs out back of fridge to collection pan beneath fridge.
- Drain Tube

# ICE OR WATER BUILDUP

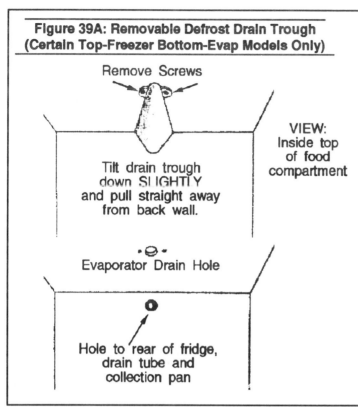

Figure 39A: Removable Defrost Drain Trough (Certain Top-Freezer Bottom-Evap Models Only)

## 6-2. DIAGNOSIS AND REPAIR

On back-evap models, if the drain backs up, it will not freeze the evaporator. You will see the collection pan full and frozen over.

On bottom-evap models, the evaporator will be frozen into one solid block of ice.

In either case, when the collection pan fills up with ice, the defrost water will generally start showing up on the inside floor of your fridge; either as water in a top freezer model or as a thick ice accretion in a side-by-side. The water may leak out onto your kitchen floor, or it may freeze up drawers within the freezer compartment.

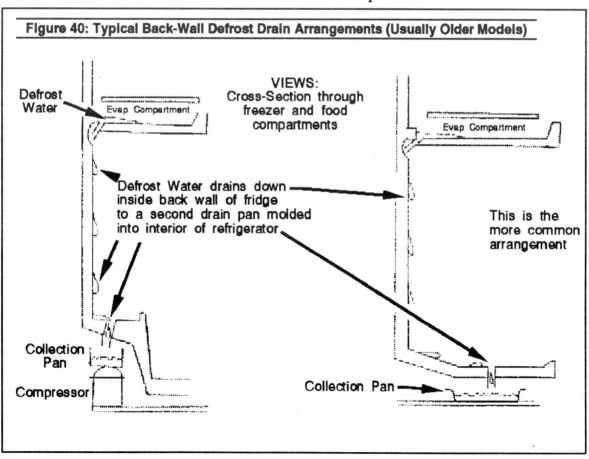

Figure 40: Typical Back-Wall Defrost Drain Arrangements (Usually Older Models)

In a top-freezer model, you will probably also see ice or drippy water on the roof of the food compartment.

Any ice or water must be removed and/or melted and the drains cleared. The fastest way to do this is to melt the ice with a blow dryer and to blow the drains clear with a pan of hot water and a syringe-type turkey baster. (Figure 41)

You will need to remove the evaporator panel to access the defrost drain. (See Section 4-2.)

Bottom-evap models are a little more difficult. The full collection pan will have frozen the evaporator into a solid block of ice, (Figure 42) which will be a bit more difficult and time-consuming to melt. *You must melt it thoroughly.* Some models have the drain hole located at the very back of the compartment instead of directly beneath the evaporator. Check there first. If the

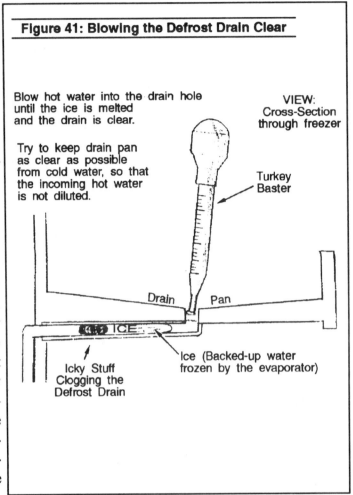

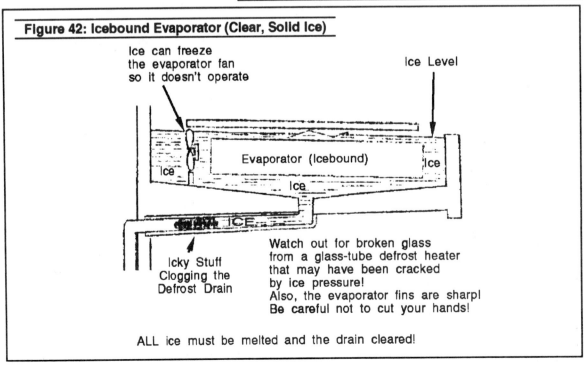

# ICE OR WATER BUILDUP

drain hole is not visible, you will need to lift the evaporator as described in Section 4-5(d) to access the drain hole.

If there is a removable drain tube or trough running along the roof of the food compartment to the back of the fridge, remove it so you can get all the ice. (Figure 39A)

As you go along, keep sucking out the excess water (now cold from melting ice) with your turkey baster and put it into an empty pan. This will prevent it from diluting the incoming hot water. It will also prevent it from ending up on your kitchen floor.

Continue blowing hot water into the drain hole until you hear it running into the drain pan beneath your fridge. Give it a few extra blasts of hot water to make sure you get all the ice. Using your turkey baster, empty your drain pan now and then. It will prevent the pan from overflowing onto the floor. If it is too inconvenient or messy to do so, don't worry about it. The water will evaporate eventually.

## 6-3. DRAIN PAN MULLION HEATERS

Some models actually have small heaters, called "mullion" heaters, attached to the underside of the collection pan. The mullion heater prevents the defrost water from re-freezing and clogging the defrost drain hole when it hits the cold defrost drain collection pan. If the mullion heater is not working properly, it may have the same effect as a clogged defrost drain.

This arrangement is generally used on back-evap fridges (top freezer or side-by-side) where the defrost heater may not be mounted close enough to the collection pan to prevent it from refreezing the drainage. These collection pan mullion heaters are a much lower wattage than the defrost heater and run a bit cooler; when operating, they will feel warm to the touch.

If you have a back-evap model, check the defrost drain pan for a mullion heater. To find it, look for its two power leads. If there *is* a mullion heater, check it for continuity and replace it if it is bad. They are usually held to the pan by spot-welded tubes or clamping plates, or by extra-sticky aluminum tape.

Re-assemble your fridge.

# Chapter 7

# *FLUKES AND UNUSUAL COMPLAINTS*

Occasionally you will run into an unusual problem that requires you to scratch your head a bit. If you know the basic systems and what they do, you can usually figure out what's going on. Following is a smattering of some of the more unusual or flukey things that I've run into as a home service technician:

## 7-1. KID CAPERS

If your fridge is warm, always check your controls *first*. Many a time, I was called to someone's house on a "warm fridge" complaint, only to find that the controls had been magically shut off.

When the little darlings of the household were queried about this divine occurrence, you could literally see the halo forming around their heads. "Did you touch the refrigerator?" Mom would ask. Standing beside the refrigerator that hadn't moved an inch in all of his (or her) six years, the child would inevitably reply: "What refrigerator, Mom?" I'm sure that my soul will burn in hell, but it was no moral dilemma for me to charge the 30 dollar service fee in such cases. One fellow whose unit I "fixed" in this way lost well over 500 dollars worth of meats from his packed-full deep freezer.

Only one time did I see a truly unexplainable occurrence of the controls getting shut off. This was a single lady of about 55 with no kids, who had owned the refrigerator for over 25 years. I'm not trying to harp on kids; it's just that as a home service tech, I've seen a lot of money wasted on unnecessary (and expensive) service calls with suspicious circumstances in homes with kids.

Do yourself a favor; check the controls first. Set them on mid-range settings. You *might* save yourself a lotta time, trouble, and expense.

## 7-2. THE HOLE-IN-THE-WALL GANG

One of the more perplexing problems that I ran into was a young couple (with 4 or 5 kids) that had a warm fridge. Upon investigation, I discovered that the Freon tubes leading to *and* from the evaporator (*above* it on their side-by-side fridge) were heavily frosted with clear, solid ice. Also, the *wires* leading to the evaporator fan and defrost heater were heavily caked with solid ice. I didn't understand *that* frost pattern at *all*, and evidently I was rushed that day, so I replaced the defrost timer and melted the ice, fully figuring I'd get a callback. Sure enough, five days later, they called me back. Same problem, same weird frost pattern.

I took off the evaporator cover panel and just stood there and STARED at the thing for about ten minutes. (Sometimes that helps me to think. You can't *do* that if the customer is standing there, looking over your shoulder. But *he* was at work, and *she* was busy with her kids.) I was just trying to think of where all that moisture could be coming from.

Suddenly, I noticed a hole, about 1" in diameter, in the upper left corner of the defrost compartment. I pulled the fridge away from the wall, and saw *daylight* through the hole. It was the pre-drilled hole for the icemaker water tube. (They did not have an icemaker, but most modern fridges come pre-drilled and wired for icemakers in case you ever want to install one.) Warm, humid air was feeding straight into the evaporator compartment!! I asked the lady about it. She said that her husband, just messing around, had pulled the little cover off the back of the fridge "about two weeks ago." I stuffed the little hole with some bubble-pack that I just happened to have around and duct-taped both sides to seal the hole.

Ignorance really *can* cost you. But at least the kids didn't do it this time.

## 7-3. DOOR SEALS AND ALIGNMENT

I personally think that door seals are one of the most misunderstood pieces of the refrigerator. Ask a do-it-yourselfer about why his fridge is warm, and the first thing he'll say after "I don't do Freon" is "but let's check the door seals." Door seals *rarely* have problems over the life of the fridge. The only ones I've seen go bad are in households that have dogs or cats that like to chew or sharpen their claws on them, or kids (there's those pesky *kids* again) that like to climb or hang on the refrigerator door. A *really* bad door seal problem is *most* likely to cause a *defrost* problem, due to humid air getting into the fridge.

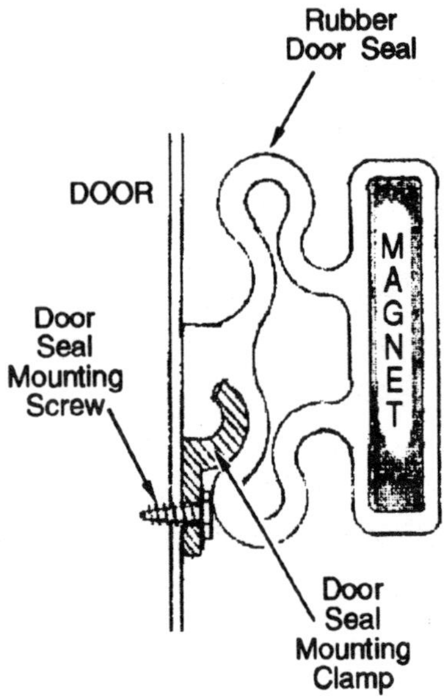

Door seals are magnetically held to the door frame. (See Figure 43 for a typical cross-section.) Unless the seal is shredded or you can physically see a gap between the seal and the door frame with the door closed, there is no reason to suspect a door seal problem.

To replace the seal, you must have a nut driver of the proper size. A power cordless drill-driver is better. A magnetic tip may prevent you from going crazy trying to hold the driver, the screw and the seal at the same time. There are lot of screws holding the seal on. Remove the screws from NO MORE THAN two sides at a time. One side at a time is better. The idea is to prevent the

**Figure 43: Typical Door Seal**

plastic inner door liner (or shelving) from drifting around--if you have to re-align it, it can be a long, frustrating, trial-and-error process. The new seal will fit in the same way as the old one came out.

You are much more likely to have a door *alignment* problem or warping. There's not much you can do with a badly warped door except to try to warp it back into shape, or replace it.

With the door closed, measure the gap around it; top and bottom, left and right. Check if the door edges and refrigerator edges are parallel. If the measurements indicate that the door is badly out of alignment, re-align it be loosening the hinges slightly (one hinge at a time) and shifting the door around. It may take a few tries to get it aligned properly.

Remove anything obstructing the seals. Sometimes the kick plate will get in the way. If it's metal, you may be able to bend it slightly to solve your problem.

## 7-4. MOVING DAY

If you have to move your fridge for any reason, make sure that you keep it upright. If you turn it on its side there's a strong probability that the compressor oil will run out of the compressor and into the condenser, and when you start the fridge, the compressor will burn out within a few hours or even minutes, for lack of lubrication. I've heard of people getting lucky and getting away with it; maybe they just *happened* to lay it on the "lucky" side of the fridge. It's not worth the risk. If your fridge has been laid on its side, stand it upright again but don't plug it in for a day or two. Just hope that either the oil drains back into the compressor or that it didn't run out in the first place.

## 7-5. A SHOCKING EXPERIENCE (MULLION HEATERS)

Once I was called to the home of a little old dog-breeder who had complained of being *shocked* while opening his refrigerator door. When I arrived, I couldn't feel anything at all--until once I happened to be leaning against the metal door of his oven (opposite his fridge) when I touched his door--ZAP! I put a volt meter between the two doors and discovered 50 volts potential!!

It turned out that the mullion (anti-sweat) heater (see section 4-1) in his door had shorted, and was grounding wherever it could--right through *him* to the oven door. Since this happened in a not-too-humid environment, the solution was simple. I disabled the heater by disconnecting and insulating the heater leads at the base of the door, where they went into the fridge. This story is made even more interesting by the fact that the little old dog-breeder's wife had just had open-heart surgery and a pacemaker installed *three weeks* before. I shudder to think what could have happened had *she* touched the fridge.

Another interesting observation about mullion heaters is that if your fridge starts to become warm, one of the first signs *may* be that the door jamb start to feel downright *hot*. This happens when the refrigerator is no longer removing heat from the doorframe and the mullion heater inside. This is especially prevalent in the doorframe between the food and freezer compartment in a side-by-side.

## 7-6. MICE CAPADES

A few years back, where I live in Southern California, we had a nasty cold spell; I mean, it was well below 45 degrees every day for a *week*. (Eat your hearts out, Midwesterners.) About a week later, I got a call from a lady with a warm fridge. Upon investigation, I found that a mouse, apparently seeking the warmth of the compressor, had crawled into there and died (whether from the heat of the compressor or from getting struck by the condenser fan, I don't know.) Anyway, he got stuck to the floor pan near the condenser fan and as he dried out, his feet and tail curled up until they finally got stuck in and stopped the condenser fan. In the home-service biz, you learn to start looking for animal and insect problems after a cold spell; lizards in the dryer vent, birds that fly into the oven roof vent and get baked and jam your oven hood exhaust fan--stuff like that.

Another neat trick that animals (especially rodents, but sometimes cats, too) like to play is to get under the fridge and start chewing wires or insulation. This can cause electrical problems, or loose insulation might get caught in the condenser fan. Cats have been known to tear off the bottom back panel of the fridge. This disturbs the airflow over the condenser and may jam the condenser fan with fiberglass insulation.

## 7-7. ICEMAKERS AND IN-DOOR WATER DISPENSERS

If you start to get an unusual buildup of ice in one particular spot in your freezer, especially beneath the icemaker, check for leaks.

Water leaking onto the kitchen floor *not* traceable to one of the problems in Chapters 4 or 6 *may* be from a leak in the icemaker or door dispenser water system. Icemakers have a water solenoid valve mounted on the back of the fridge, usually behind the back bottom panel, and a water tube that leads straight up the back of the fridge to the icemaker. Door dispensers have a similar water solenoid valve with a water tube that leads beneath the fridge and into the door through a hollow hinge. If your fridge has *both* features, it's usually a dual water solenoid valve. It *can* be simple or not-so-simple to fix, but basically it's just a plumbing job.

## 7-8. SEE THE LIGHT

On several occasions, I have been called to peoples' homes with the complaint that the freezer or food section was cold in the bottom, but felt like a warming oven was on in the top of the compartment.

It turned out in every case that the interior lights were not shutting off. (A light bulb puts out enough heat to actually *warm* the top of the compartment. Remember, warm air rises.) There are two things that might cause this.

One is a defective door switch, easily diagnosed and corrected.

The other is if nothing is *contacting* the door switch. Some units were built with a removable interior shelf that also shut off the light by contacting the door switch. Remove the shelf, and nothing hits the switch; the light doesn't turn off.

The easiest way to diagnose this is to peek into the compartment while slowly closing the door. If the light does not shut off well before the door is fully closed, test the switch and look at its closing mechanism and see what's happening.

## 7-9. BAD ODORS

To my experience, this is one of the toughest problems to solve. I have been called out on a variety of "bad odor" complaints, and rarely are they solvable. Modern fridges, with few exceptions, have plastic interiors that *do* absorb odors to some degree, rather than the porcelain interiors standard in days of yore. If you have an odor problem, it *may* help to keep an open box of baking soda in there, or to wash out the interior of the fridge with a very mild bleach solution.

If your fridge has had a meltdown (thawed out with everything in it), try pulling off whatever panels you can and see if any meat blood or other smelly stuff has gotten under them. Styrofoam panels can absorb odors, too--but try washing with a mild soap solution.

One semi-solvable odor problem that I ran into was a fellow whose fire-engine red fridge was getting warm *and* it had an absolutely acrid smell within. It turned out that the defrost heater lead within the defrost compartment had shorted out and burned off a bunch of its heavy rubber insulation, then melted itself so the defrost system was not working. I fixed the fridge, but I do not know if he ever got out the burnt-rubber smell.

## 7-10. FIX THE LITTLE STUFF

Have you had a broken door catch or handle? For *years*? A light bulb out? How about the kickplate--how long has *it* been falling off? Maybe you have a missing shelf, or one shelf that's cracked. *You* know, every time you put something on that shelf, you have to do a quick mental weight and stress calculation to make sure everything on it doesn't end up on the kitchen floor. It's *really* annoying to think about, isn't it? So naturally, you just stop thinking about it.

Take the time to make it right! Plastic parts are usually not that expensive; though things like shelves usually must be gotten through the dealer or factory warehouse. Special appliance light bulbs are a couple of bucks at your nearest appliance parts dealer. How much is your peace of mind worth?

## 7-11 STRANGE NOISES

Actually, there aren't too many things in a fridge that can cause a lot of noise. A few that you might commonly hear are:

*A* **rattling or a buzzing noise** might be coming from the evaporator or condenser fan hitting something. It may also be that the defrost drain pan or compressor mounts are loose and something is vibrating a bit.

*A* **whistling or warbling sound** usually is coming from the evaporator fan motor. Replace as described in section 4-3.

*A* **hissing or gurgling sound** might happen for a few minutes after the compressor shuts off. This is the Freon flowing through the tubing in the system. When the pressure throughout the system equalizes, the noise will stop. If there is no evaporator fan running, you *may* hear the Freon gurgling through the evaporator at any time the fridge is running.

## 7-12. FIRE IN THE FRIDGE!!!

Okay, *you're* the tech. You get a frantic call from a customer saying there's a red glow in the bottom of the freezer in his side-by-side; he's afraid that something's burning. He doesn't hear it running, but he does hear popping and hissing noises. You rush over to his home, and when you get there, you don't see any red glow. You *do* notice that the light doesn't work in the freezer. He says that it went out months ago, and he hasn't gotten around to replacing it. He has had the fridge for ten years, and has never seen a red glow down there before. What do you do?

If you're me, you grab a screwdriver, reach down to the defrost timer and set it on "defrost" again. Then you reach into your tool kit for your trusty, dusty extendable inspection mirror. You show the customer, via the mirror, the red-glowing defrost heater. You explain the defrost system and why he's never noticed a red glow before:

When the light bulb was working, the interior of the fridge was *so* bright that he just couldn't *see* that dim red glow. And the defrost heater is only *on* for about 10 or 15 minutes every 6 to 8 hours. What are the odds against his opening the fridge at *just* the right time and seeing the fridge in a defrost cycle?

Then you sell him a special appliance light bulb for three-fifty, plus a second one as a spare, and you charge him a 30-dollar service call fee besides.

As I said before, ignorance *truly can* cost you.

# Index

## A
Air Doors 2, 15-16
Alligator Jumpers 8, 51
Ammeter 8, 11
Ammonia System 3

## B
Back Bottom Panel 13, 19-20, 64
Blow Dryer 8, 13, 58
Butter Conditioners 17

## C
Capillary Tube 46
Cam, Drive, *see* Drive Cam
"Cheater" Wire *see* Alligator Jumpers
Chill-Type Refrigerators 3
    Chill Plates 3, 26
    Chill Tubes 3, 26
Cold Control 2, 15-16, 45-46, 51
Compressor 2, 13, 15, 49, 51-53, 63, 76
    Testing For Power, 51
    Starting Components 52-53
Condenser 2, 17-21, 63
    Cleaning, 18-19
Condenser Brush 8, 18-19
Condenser Fan 2, 17-21, 65-66
    Motor, 20-21
Continuity, Testing 10-11
Controls 15-17, 49, 61

## D
Defrost Drain System 2, 55-59
Defrost Heater 2, 30-33, 40, 66
    Diagnosing, 36-37
    Replacing, 37-38
Defrost Problems 13, 24
Defrost Switch, Whirlpool / Kenmore Flex-Tray Icemaker 40-41, 49-50
Defrost System 1-2, 28-29, 35-40,
    Hot Gas Defrost System 3, 44

Defrost Timer 2, 28, 49
    Advancing, 35
    Replacing, 36
Door Alignment 63
Door Seals 62-63
Door Switches 22, 64-65
Drain Trough, *see* Defrost Drain System
Drain Tube, *see* Defrost Drain System
Drain Pan 66, *see also* Defrost Drain System
Drive Cam, Whirlpool / Kenmore Flex-Tray Icemaker 40-42
Drive Gears, Whirlpool / Kenmore Flex-Tray Icemaker 40-42
Drive Motor, Whirlpool / Kenmore Flex-Tray Icemaker 40-41

## E
Electricity 9, 52
"Energy Saver" Control, 17
Evaporator 1, 26-27
    Icebound 57-59
    Lifting 37-38
Evaporator Access Panel 25-26, 27
Evaporator Fan 1, 21-23, 65-66
    Motor 22-23, 66

## F
Freon 1, 2, 44, 45, 66
Frost 1, 13, 24, 27-28, 44, 61-62
    Pattern and Quality, 27-28, 45
Flex-Tray Icemaker, *see* Whirlpool / Kenmore Flex-Tray Icemaker

## G
Gas Refrigerators 3
Gears, Drive, Whirlpool / Kenmore Flex-Tray Icemaker, *see* Drive Gears
Gears, Timing, Whirlpool / Kenmore Flex-Tray Icemaker, *see* Timing Gears

## H

Heat Flow 1
Heater, *see* Defrost Heater
 *see also* Mullion Heater
Hot Gas Defrost System *see* Defrost System

## I

Ice 12-13, 24, 27-28, 55-59, 61-62
Icemaker 28, 38-43, 64
Inspection Mirror 9
Insulation 13, 26, 64

## L

Light Bulb 64-65, 66
Lubrication, of Motors, Timers, Etc. 13

## M

Manual Defrost 3
Meat Compartment Controls 17
Mirror, Inspection 9
Moving Your Refrigerator 63
Mullion Heater 17, 59, 63-64

## N

Nameplate Information 5-6
Noises 13, 22, 49, 52-53, 65-66

## O

Odors 65
Oil, Compressor 63
Oil, Lubricating *see* Lubrication
 *see also* Moving Your Refrigerator

## P

Parts Dealer 5
Parts 5-7

## R

Repair and Safety Precautions 12-13

## S

Safety Precautions 12-13
Seals, Door *see* Door Seals
Sounds, *see* Noises
Sweat 17

## T

Temperature 1
Terminating Thermostat 2, 34, 36-37
Thermostat, *see* Cold Control
 *see also* Terminating Thermostat
Three-In-One 52-53
Timer, see Defrost Timer
Timing Gears, Whirlpool / Kenmore Flex-Tray Icemaker 41
Tools 7
Turkey Baster 8, 58-59

## V

Vegetable Compartment Controls 17
VOM *see* Volt-Ohmmeter
Volt-Ohmmeter 8, 9-11, 51-52
Voltage Testing 9, 22, 51-52

## W

Water Buildup 55-59, 64
Water Dispensers 64
Whirlpool / Kenmore Flex-Tray Icemaker 28, 38-43, 49-50
Wiring 51-52, 64

## ABOUT THE AUTHOR

Douglas Emley is Chief Officer in charge of hazardous materials on-board a Merchant Marine ship. He holds a Bachelor of Science degree, engineer's license and officer's license from the Kings Point Merchant Marine Academy, Long Island, New York. Emley has been a major appliance service technician for nearly ten years. Tired of seeing individuals pay his service fees for simple repairs, Emley decided to write easy-to-understand repair guides. "The manufacturers service manuals are too confusing for the average do-it-yourselfer. In my opinion, they contain far too much unnecessary information."

To simplify the diagnosis and repair process, Emley deliberately avoids technical terminology in his instructions. "I'd rather show the average person how to save a fortune by diagnosing the problem themselves and fixing the simple stuff -- 95% of all repairs! When there's a *serious* problem, then call out the tech and pay for his expertise."

If you found this guide easy-to-follow and helpful, you may be interested in Emley's other repair guides:

### The No Headache Guide to Home Repair Series--
**Washing Machine Repair Under $40** -- ISBN 1-884348-02-5
**Clothes Dryer Repair Under $40** -- ISBN 1-884348-01-7
**Dishwasher Repair Under $40** -- ISBN 1-884348-03-3

These books can be ordered through your local library, bookstore or by mail order. Send check or money order to: New Century Publishing, P.O. Box 9861, Fountain Valley, CA 92708. Each title is $12.95; please include $2.50 shipping via book rate (allow 4 to 5 weeks delivery) or $4.50 shipping via Air Mail. California residents please add 7.75% sales tax. Volume discounts are available by calling (714)554-2020.